AF361654

Life Cycle & Technoeconomic Modeling

Life Cycle & Technoeconomic Modeling

Editors

Antonio Colmenar Santos
David Borge Diez
Enrique Rosales Ascnsio

MDPI • Basel • Beijing • Wuhan • Barcelona • Belgrade • Manchester • Tokyo • Cluj • Tianjin

Editors

Antonio Colmenar Santos
National University of Distance Education
Spain

David Borge Diez
University of Léon
Spain

Enrique Rosales Asensio
University of La Laguna
Spain

Editorial Office
MDPI
St. Alban-Anlage 66
4052 Basel, Switzerland

This is a reprint of articles from the Special Issue published online in the open access journal *Energies* (ISSN 1996-1073) (available at: https://www.mdpi.com/journal/energies/special_issues/life_cycle_technoeconomic_modeling).

For citation purposes, cite each article independently as indicated on the article page online and as indicated below:

LastName, A.A.; LastName, B.B.; LastName, C.C. Article Title. *Journal Name* **Year**, *Article Number*, Page Range.

ISBN 978-3-03943-639-2 (Hbk)
ISBN 978-3-03943-640-8 (PDF)

© 2020 by the authors. Articles in this book are Open Access and distributed under the Creative Commons Attribution (CC BY) license, which allows users to download, copy and build upon published articles, as long as the author and publisher are properly credited, which ensures maximum dissemination and a wider impact of our publications.

The book as a whole is distributed by MDPI under the terms and conditions of the Creative Commons license CC BY-NC-ND.

Contents

About the Editors

Antonio Colmenar Santos has served as Senior Lecturer in the field of Electrical Engineering at the Department of Electrical, Electronic and Control Engineering at the National Distance Education University (UNED) since June 2014. Dr. Colmenar-Santos was previously Adjunct Lecturer at both the Department of Electronic Technology at the University of Alcalá and the Department of Electric, Electronic and Control Engineering at UNED. He has also worked as a consultant for the INTECNA project (Nicaragua). He has been part of the Spanish section of the International Solar Energy Society (ISES) and of the Association for the Advancement of Computing in Education (AACE), working in a number of projects related to renewable energies and multimedia systems applied to teaching. He was the coordinator of both the virtualization and telematic services at ETSII-UNED, and Deputy Head Teacher and Head of the Department of Electrical, Electronics and Control Engineering at UNED. He is the author of more than 60 papers published in respected journals (http://goo.gl/YqvYLk) and has participated in more than 100 national and international conferences.

David Borge Diez has a Ph.D. in Industrial Engineering and an M.Sc. in Industrial Engineering, both from the School of Industrial Engineering at the National Distance Education University (UNED). He is currently Lecturer and Researcher at the Department of Electrical, Systems and Control Engineering at the University of León, Spain. He has been involved in many national and international research projects investigating energy efficiency and renewable energies. He has also worked in Spanish and international engineering companies in the field of energy efficiency and renewable energy for over eight years. He has authored more than 40 publications in international peer-reviewed research journals and participated in numerous international conferences.

Enrique Rosales Asensio (Ph.D.) is an industrial engineer with postgraduate degrees in Electrical Engineering, Business Administration, and Quality, Health, Safety and Environment Management Systems. He has served as Lecturer at the Department of Electrical, Systems and Control Engineering at the University of León, and Senior Researcher at the University of La Laguna, where he has been involved in a water desalination project in which the resulting surplus electricity and water would be sold. He has also worked as a plant engineer for a company that focuses on the design, development, and manufacture of waste-heat-recovery technology for large reciprocating engines, and as a project manager in a world-leading research center. He is currently Associate Professor at the Department of Electrical Engineering at the University of Las Palmas de Gran Canaria.

Preface to "Life Cycle & Technoeconomic Modeling"

This book aims to perform an impartial analysis to evaluate the implications of the environmental costs and impacts of a wide range of technologies and energy strategies. This information is intended to be used to support decision-making by groups, including researchers, industry, regulators, and policy-makers. Life cycle assessment (LCA) and technoeconomic analysis can be applied to a wide variety of technologies and energy strategies, both established and emerging. LCA is a method used to evaluate the possible environmental impacts of a product, material, process, or activity. It assesses the environmental impact throughout the life cycle of a system, from the acquisition of materials to the manufacture, use, and final disposal of a product. Technoeconomic analysis refers to cost evaluations, including production cost and life cycle cost. Often, in order to carry out technoeconomic analysis, researchers are required to obtain data on the performance of new technologies that operate on a very small scale in order to subsequently design configurations on a commercial scale and estimate the costs of such expansions. The results of the developed models help identify possible market applications and provide an estimate of long-term impacts. These methods, together with other forms of decision analysis, are very useful in the development and improvement of energy objectives, since they will serve to compare different decisions, evaluating their political and economic feasibility and providing guidance on potential financial and technological risks.

Antonio Colmenar Santos, David Borge Diez, Enrique Rosales Asensio
Editors

Article

Increasing the Competitiveness of Tidal Systems by Means of the Improvement of Installation and Maintenance Maneuvers in First Generation Tidal Energy Converters—An Economic Argumentation

Eva Segura [1], Rafael Morales [1,*] and José A. Somolinos [2]

[1] Escuela Técnica Superior de Ingenieros Industriales de Albacete, Universidad de Castilla-La Mancha, 02071 Albacete, Spain

[2] Escuela Técnica Superior de Ingenieros Navales, Universidad Politécnica de Madrid, 28040 Madrid, Spain

* Correspondence: Rafael.Morales@uclm.es; Tel.: +34-967-599-200 (ext. 2542); Fax: +34-967-599-224

Received: 23 April 2019; Accepted: 18 June 2019; Published: 26 June 2019

Abstract: The most important technological advances in tidal systems are currently taking place in first generation tidal energy converters (TECs), which are installed in areas in which the depth does not exceed 40 m. Some of these devices are fixed to the seabed and it is, therefore, necessary to have special high performance ships to transport them from the base port to the tidal farm and to subsequently recover the main units of these devices. These ships are very costly, thus making the installation costs very high and, in some cases, probably unfeasible. According to what has occurred to date, the costs of the installation and maintenance procedures depend, to a great extent, on the reliability and accessibility of the devices. One of the possible solutions as regards increasing system performance and decreasing the costs of the installation and maintenance procedures is the definition of automated maneuvers, which will consequently influence: (i) an increase in the competitiveness of these technologies; (ii) a reduction in the number and duration of installation and maintenance operations; (iii) less human intervention, or (iv) the possibility of using cheaper general purpose ships rather than high cost special vessels for maintenance purposes, among others. In this research, we propose a definition of the procedures required for the manual and automated installation and maintenance maneuvers of gravity-based first generation TECs. This definition will allow us to quantify the costs of both the manual and automated operations in a more accurate manner and enable us to determine the reduction in the cost of the automated installation and maintenance procedures. It will also enable us to demonstrate that the automation of these maneuvers may be an interesting solution by which to improve the competitiveness of tidal systems in the near future.

Keywords: ocean energy; tidal energy converters; offshore renewable energy; life-cycle costs; installation and maintenance maneuvers; economic-financial viability

---

1. Introduction

The large-scale exploitation of fossil fuels has had important environmental repercussions such as climate change or the rise in sea level amongst others [1,2]. This means that there is currently a need to reduce the dependence on fossil fuels and place greater emphasis on renewable energy sources in order to fulfil future sustainable energy needs [3,4]. In 2009, the European Union (EU) established that 20% of final energy consumption should originate from renewable sources by 2020 [5], and additionally set 2050 as the target year by which emissions will have been reduced by 80% [6,7]. In this context, and with the aim of achieving a sustainable development, in 2015, the United Nations Framework Convention on Climate Change was established in order to reduce the causes of climate change as regards food production and limit the increase in temperature (increases of up to 1.5 °C) [8]. A number of viable renewable energies could, therefore, be exploited to achieve this goal, among which marine renewable energy (MRE) is attracting increased attention [9].

MRE is currently recognized as being an abundant, geographically diverse energy resource that has both the public's acceptance and positive associated externalities (economic growth, job creation or the mitigation of the negative impacts of climate change, etc.) [10]. It can be exploited from offshore wind, waves, tides, tidal currents, thermal gradients or salinity gradients. This paper is focused on the exploitation of the tidal current resource, which it is hoped will play a major role in meeting future energy needs with regard to other renewable energy sources thanks to its high predictability, stability and high load factor [11]. If we wish to employ technologies to harness the energy obtained from tidal currents in order to attain sustainable development, it is necessary to use natural resources in an efficient manner, i.e., we must optimize their exploitation [12]. Devices that can be utilized to harness tidal current power where the depth is no greater than 40 m have been developed by various technology manufacturers [13,14]. These devices, which are denominated as first generation tidal energy converters (TECs), are normally supported on bases that are fastened to the seabed by means of various types of anchoring systems (monopole, piloted or gravity). It is undoubtedly technically feasible to employ these devices to harness energy from tidal currents, although very few tidal stream projects are currently operating at a commercial stage [15]. This is principally owing to the fact that it typically costs more to generate energy from tidal currents than it does when using other renewable technologies [16]. It is consequently vital to understand the parameters that may affect the cost structure in order to provide a framework containing areas in which these costs can be reduced [17].

A detailed analysis of the life cycle costs (LCC) for first generation tidal energy farms (TEFs) shows that the installation and maintenance procedures are of the utmost importance and must be optimized in order to increase their current potential, help their acceleration and sustainability and help attract investment in these technologies [18,19]. These procedures include the transportation of each of the TECs from the base port to its installation site, the preparation of the seabed, the placement and installation of anchor systems and/or the deployment of a mooring system, and the positioning, connection and disconnection of the main units of the devices [20,21]. These tasks necessitate the use of special high-performance ships that are equipped with dynamic positioning, large cranes, etc., and this implies high installation and maintenance costs [22,23]. The installation and maintenance costs are greatly dependent upon the accessibility and reliability of the device [16]. It would, however, be possible to increase system performance and decrease the aforementioned costs by automating the performance of the immersion and emersion maneuvers [24,25]. This can be done by controlling the ballast water inside the devices, which consequently permits the implementation of a closed loop depth and/or orientation control that makes(s) it possible to: (i) raise the generation unit from the seabed to the surface of the sea and (ii) carry out the same operation but in reverse. We can perform these automatic maneuvers by using small guide wires and by controlling the ballast water inside the device. The achievement of this objective will consequently influence: (i) less human intervention, (ii) the possibility of using the cheapest general purpose ships rather than high cost special vessels for maintenance purposes, (iii) a reduction in the number and duration of installation and maintenance operations or (iv) an increase in the competitiveness of these technologies, among others. The potential benefits of these systems are very important, but, as they are in an early stage of development, studies that address the economic feasibility of these systems have not yet been developed. Several authors have produced interesting papers comprising feasibility studies concerning other types of offshore projects. These include: wind energy [26], wave energy [27], co-located projects (wind and wave energy) [28] and hybrid projects (wind and wave energy) [29]. No economic-financial studies focusing on the automation of installation and maintenance maneuvers have, however, been produced to date.

The main contributions of this research are the following: (i) we discuss the merits of automated installation and maintenance maneuvers with regard to manual maneuvers for an idealized gravity (a substantial mass is used to support the structure on which the TEC is placed.) -based first generation TEC designed by our research group (Grupo de Investigación Tecnológico en Energías Renovables

Marinas, *GIT-ERM*); (ii) we provide interesting information about manual and automated installation and operation maneuvers for these tidal energy technologies, which is not usually found in scientific literature as these technologies are at an initial (pre-commercial) stage of development, and (iii) we carry out a comparative economic-financial feasibility study for these maneuvers, which illustrates that the development of advanced automation systems for these maneuvers may be a very interesting approach by which to increase the competitiveness of this source of renewable energy in the near future.

The remainder of the paper is organized as follows: Section 2 describes the procedures used to carry out installation and maintenance maneuvers for first generation TECs in both a manual and an automated fashion. The procedure used to evaluate the economic-financial feasibility of tidal energy projects using manual or automated maneuvers is briefly explained in Section 3. Section 4 shows the results attained after carrying out a numerical case study of a 50 MW TEF in the cases of both manual and automated maneuvers. Finally, Section 5 is devoted to our conclusions and proposals for future works.

2. Description of Installation and Maintenance Maneuvers for Gravity-Based First Generation TECs

In this section, we provide details on the installation and maintenance maneuvers for gravity-based first generation TECs. It should be noted that the information regarding these sorts of maneuvers is very limited owing to the fact that these technologies are currently at an initial stage of development (pre-commercial stage), signifying that real data about these maneuvers is not yet available [30]. The implementation of improved procedures for installation and maintenance maneuvers will actively influence their successful future commercialization [31], and this is one of the most important aspects studied by the GIT-ERM research group. The vessels used to perform these maneuvers should have the following characteristics: (i) *Dynamic positioning*, which allows redundancy in order to ensure work under extreme conditions and to guarantee security and reliability while these maneuvers are being carried out. These kinds of vessels have a high level of technology and are very costly to acquire/rent [32,33]; (ii) a *Heavy lifting crane*. Any cranes operating with these gravity-based first generation TECs must have a lifting capacity of around 250 tons [34], and (iii) the special vessel needs to have a high area on its deck on which to transport the structure, gondolas, auxiliary tools, etc. The aforementioned considerations allow us to conclude that the characteristics of the vessels required to carry out these maneuvers are not typical since the number of specialized vessels is not currently high and they are not easy to find on the market. They are, at present, used in the installation and maintenance of offshore wind energy farms and in the oil and gas industry, but the cost of hiring them is currently very high and oscillates according to the market (thus causing a high economic dependence). Furthermore, other sorts of vessels, such as remotely operated vehicles (ROVs), cable-laying vessels or tug vessels, among others, will be necessary to provide these special vessels with support when performing the installation and maintenance maneuvers, and these are very costly [35]. The following subsections deal with the definition of the installation and maintenance maneuver methods for gravity-based first generation TECs using, in the first case, manual and, in the second case, automated control. However, before performing the TEC installation and maintenance maneuvers, several stages have to be carried out on the TEF, which are graphically illustrated in Figures 1 and 2 and explained below:

- *Installation sequence at the tidal farm level*: The first elements to be installed are the transformation platform and the converters. Bearing in mind the depth and the composition of the seabed on which the TEF is installed (around 40 m), the use of a jacket platform is recommended owing to the fact that it is very safe, in addition to being highly adaptable and reliable [36,37]. The following element to be installed is the exportation cable, which requires the use of a cable-laying vessel (Figure 1a). The cable-laying vessel transports the umbilical cable from the transformation platform to the special vessel in charge of transporting the TEC (Figure 1b), and the connection between the base structure and the transformation platform is, therefore, achieved (Figure 1c). The cable-laying vessel waits until the base TEC has been installed (Figure 1d), after which it is possible to install the base structure on the seabed by means of gravity (the procedure employed to install the base structure will be described below and is illustrated in Figure 2). Once the base support has been installed, the cable is extended in order to connect it to the adjacent TEC. During this procedure, the installation vessel has sufficient time to return to the base port and then return to the TEF with a new device. The cable-laying vessel waits to be given the end of the cable in order to perform the connection between the end of that cable and the new TEC structure and to repeat the cable connection process that will join it to the next TEC. This process is repeated until the TEF is completely installed.

- *Installation of the submarine cables*: It should be noted that it is fundamental to provide the interconnection cables and the exportation cables with adequate protection in order to avoid possible natural damage (resulting from earthquakes or movements caused by waves and currents) or damage caused by human activities (anchors or fishing artifacts, among others). The protection usually employed is that of burying the cables to a sufficient depth (from 0.5 m to 1 m) [38]. The following equipment is required to install the submarine cables: (i) a cable-laying vessel with its auxiliary equipment; (ii) ROVs to perform the trenching and burial processes; (iii) tug vessels with cranes and a diving team; and (iv) ground equipment, such as excavators, winches, trucks, etc. The procedure employed is the following: the cable-laying vessel is in charge of depositing the cables on the seabed following the most homogeneous path in order to avoid zones with rocks (Figure 2a). The trenching process is carried out in the opposite direction to the cable-laying process and is performed by a ROV-trencher (Figure 2b). This device is in charge of removing the cable, making the trench and placing the cable inside the trench [39]. The same ROV (but using a different tool) then performs the burying process in the opposite manner to the trenching process, thus leaving the cable completely covered (Figure 2c,d).

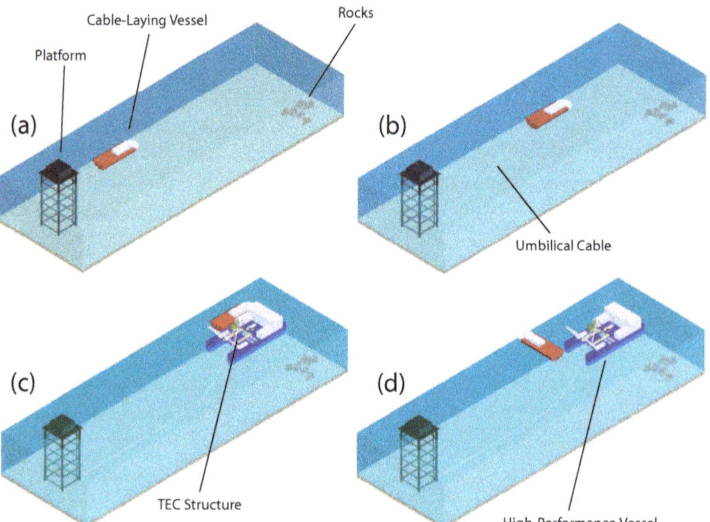

Figure 1. Installation sequence at tidal farm level: (**a**) joining the umbilical cable to the platform by means of a cable-laying vessel; (**b**) umbilical cable-laying process; (**c**) connection of the umbilical cable to the TEC (Tidal Energy Converter) structure; and (**d**) cable-laying process of the next umbilical cable.

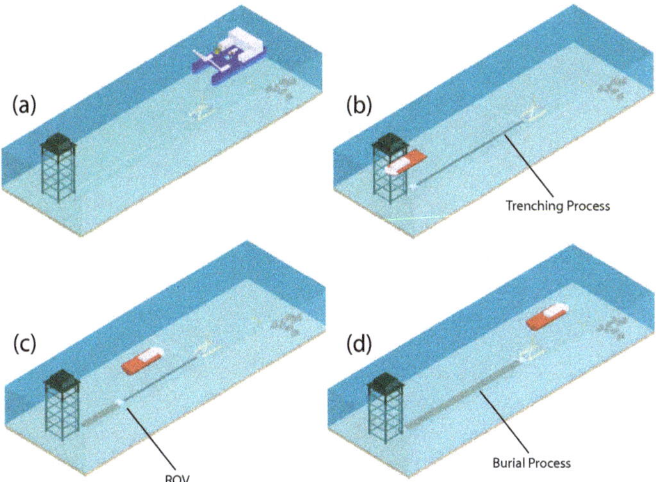

Figure 2. Installation of the submarine cables: (**a**) installation of the base; (**b**) ROV (Remotely Operated Vehicle) performing the trenching process of the first umbilical cable; (**c**) ROV starting umbilical cable burial process; and (**d**) ROV finishing the burying process of the umbilical cable.

2.1. Manual Installation and Maintenance Maneuvers for First Generation TECs

We define the term *manual* for installation and maintenance maneuvers for TECs in an open loop. These sorts of maneuvers are currently used in the first generation tidal technologies which are, at present, in a pre-commercial stage [38]. An example of the gravity-based TEC described in this section and designed by the GIT-ERM research group is illustrated in Figure 3. The manual installation of the TECs can be divided into the installation sequence for the support structure of the TECs and the installation sequence for the gondolas. The following steps have, therefore, been defined in order to perform the installation tasks:

Figure 3. Example of TEC used for manual installation and maintenance maneuvers. Additional details about this TEC design can be found in [38].

- *Installation of the support structure of the TECs*:

    - The special vessel transports the complete TEC and the equipment required (support structure, ballasts, gondola, etc.) simultaneously (see Figure 4a), and moves from the base port towards the TEF. When it is at the TEF, it uses its dynamic positioning system to place all the necessary items in the exact position in which the TEC will be installed.

- The umbilical cables are then connected to the base structure and the guide cables used to recover the gondola are attached to the deck of the vessel, thus preparing the TEC structure for its installation (see Figure 4b).
- The crane on the special vessel raises the TEC structure off the special vessel by means of four cables, and the descent process begins. The descent process is performed thanks to the weight of the TEC structure and the guide cables, and the descent velocity and the orientation of the TEC structure with regard to the special vessel are controlled (see Figure 4c). When the structure is correctly positioned, special concrete bags are released in order to fix the TEC structure to the seabed (see Figure 4d). The cables used during the descent process are subsequently removed.
- The ballasts are placed on the TEC structure. This operation is performed by the crane, and the ballasts are lowered one by one (see Figure 4e).
- Finally, the guide cables are detached from the vessel and are submerged by means of a ballast and a buoy in order to recover them during the future gondola installation process. These cables are placed on the seabed in a zone that does not involve risks as regards the installation procedures of the other devices, the farm or the umbilical cables. The TEC structure is now considered to be completely installed (see Figure 4f).

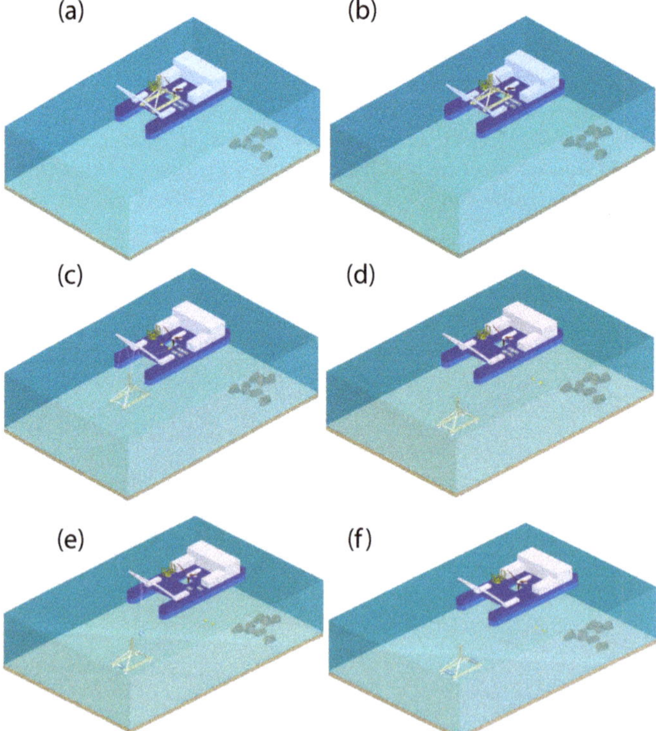

Figure 4. Installation of the structure of the TECs: (**a**) position required to install the base; (**b**) connection of the umbilical cables to the TEC structure; (**c**) controlled descent of the TEC structure; (**d**) fixing the TEC structure to the seabed; (**e**) placement of the concrete ballasts; and (**f**) installation of the TEC structure once the process has been completed.

- *Installation of the gondolas:*

  - Once the support structure has been completely installed, the special vessel is placed on the TEC structure and the guide cables of the gondola are recovered by means of an acoustic signal (see Figure 5a).
  - In order to work with the gondolas, a specific tool equipped with a hydraulic system will be used, whose objective is to wrap itself around the gondola that is to be installed or recovered. Its operation is similar to that of a clamp (see Figure 6).
  - The guide cables are connected to the tool used to lower the gondola (see Figure 5b). These cables facilitate the descent of the gondola and the insertion of the gondola into the structure.
  - The gondola initiates its descent with the guide cables thanks to its own weight and without oscillations until the gondola has been inserted into the structure. Figure 5c illustrates the descent process of the gondola and Figure 5d depicts the gondola-structure insertion process.
  - The final step is that of removing the tool used to install the gondola and the retrieval of the guide cables. Figure 5e illustrates the removal process. When the tool is on the deck of the vessel, the guide cables are removed from the tool and are submerged in a safe location by means of a ballast and a buoy in order to recover them during the next intervention. Figure 5f shows the installation of the whole TEC once the process has been completed.

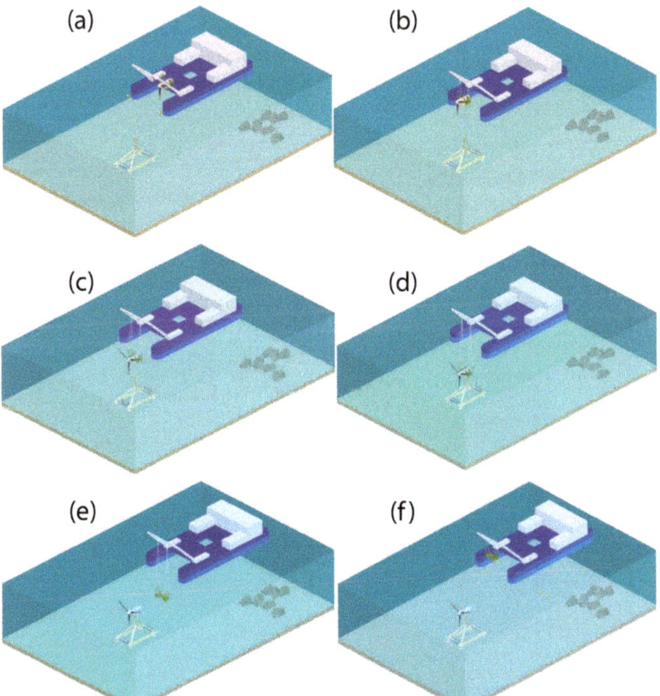

Figure 5. Installation of the gondola of the TECs: (**a**) cable-recovery process; (**b**) connection of the cables to the tool in charge of lowering the gondola; (**c**) controlled descent of the gondola; (**d**) process of inserting the gondola into the TEC structure; (**e**) tool and cable removal process; and (**f**) end of gondola-installation process.

Figure 6. Tool used for manual installation and maintenance maneuvers (installation and recovery of the gondola).

The maneuvers that are necessary to perform the maintenance tasks (recovery of a submerged gondola) follow the inverse order of that described for the installation of the gondola. The procedure for maintenance maneuvers, therefore, shares a lot of similarities with the procedure of installing the gondolas, and the steps required to perform maintenance maneuvers are the following:

- *Recovery of a submerged gondola*:

  - The starting point is that of locating the special vessel above the gondola to be recovered. The first step is the recovery of the cables from the seabed. The ends of the cables are released from the seabed by means of an acoustic signal (see Figure 7a) and these cables are connected to the tool used to recover the gondola.
  - The tool starts its descent, following a trajectory with an inclination angle that permits the tool to wrap itself around the back of the gondola (see Figure 7b).
  - When the tool is ready to perform the grip, the cables are tightened and placed completely vertically (see Figure 7c). The hydraulic system of the tool is activated in order to close it and fix it to the gondola (see Figure 7d).
  - The process of raising the gondola begins. As the cables are tightened, the displacements are very small and the operation is carried out under safe conditions (see Figure 7e).
  - When the whole system (gondola + tool) is outside the water, the cables are removed from the tool and are submerged again by means of a ballast and a buoy in order to recover them in the future (see Figure 7f).

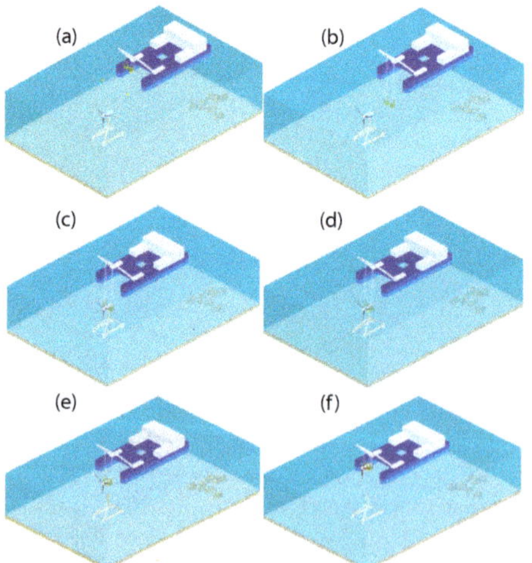

Figure 7. Maintenance operations of the gondola: (**a**) positioning the special vessel above the gondola; (**b**) descent process of the tool; (**c**) coupling process between the tool and the gondola and cable tensioning process; (**d**) activation of the hydraulic system of the tool to fix it to the gondola; (**e**) gondola lifting process; and (**f**) gondola recovered.

2.2. Automated Installation and Maintenance Maneuvers for First Generation TECs

Section 2.1 shows that the manual installation and the maintenance maneuvers are both very complex and very costly. Moreover, the zones in which tidal energy technologies operate economically are zones with peak tidal velocities greater than 2.5 m/s [40,41] and are also characterized by the fact that they are zones with adverse climatologic conditions that increase the complexity of these maneuvers [42]. The development of automated installation and maintenance maneuvers, which help to reduce the resources required, the complexity of the operations and the costs are, therefore, a very interesting point to research. We define the term *automated* for TEC installation and maintenance maneuvers in a closed loop. These sorts of maneuvers have recently been presented by the GIT-ERM research group through several recent patents [43–45] as a solution that will influence tidal energy systems in the following particular aspects [22,23]: (a) the number and duration of the installation operations will be reduced; (b) the profitability of the project will be increased; (c) there will be less human intervention; (d) the weather window will be maximized, and (e) it may be possible to employ general-purpose ships as tugboats for maintenance purposes, rather than high-cost specialist vessels. In the following subsections, we provide details on the modifications developed by the GIT-ERM research group and made to the TEC proposed in Figure 3 in order to perform automated immersion and emersion maneuvers, along with the definition of the procedures employed to install and maintain these advanced systems

2.2.1. Modifications Made to TECs in Order to Perform Automated Maneuvers

The GIT-ERM research group designed the gravity-based first generation TEC presented herein in order to enable it to perform automatic emersion/immersion maneuvers. This is done by using small guide wires and controlling the ballast water inside the device [24,25]. The control of the ballast water permits the implementation of a closed-loop depth and/or orientation control, which, in turn, allow(s): (i) the extraction of the main power generation unit from its normal depth of operation (on the seabed) to the surface of the sea, and (ii) it to be returned from the surface to its base on the seabed. Figure 8 illustrates the shape of a gravity-based first generation TEC capable of performing automated maneuvers and its distribution equipment. The main differences between the gondola of the TEC illustrated in Figure 3 (designed for manual maneuvers) and the gondola of the TEC depicted in Figure 8 (designed for automated maneuvers) are the following: (a) the places in which the ballast tanks and their associated pumping system are located, and (b) the shape of the gondola, which has been modified in order to attain neutral buoyancy when the ballast tanks are half full. In the system depicted in Figure 8a, the gondola has been increased longitudinally as opposed to increasing its diameter. This has been done so as to optimize its hydrodynamic performance [37]. A detailed description of the design modifications and the behavior of the modified gravity-based TEC, along with interesting laboratory experiments, can be found in [23–25,37].

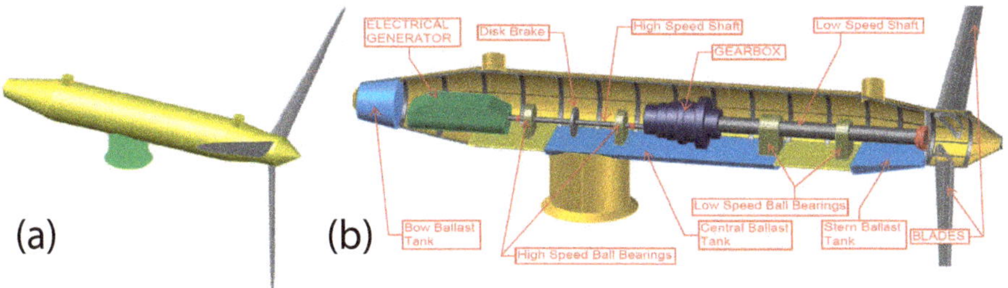

Figure 8. First generation TEC designed for automated maneuvers: (**a**) shape of the gondola and (**b**) distribution equipment.

2.2.2. Automated Installation and Maintenance Maneuvers

In this section, we propose a procedure with which to carry out the automated installation and maintenance maneuvers for gravity-based first generation TECs by controlling the inner ballast water inside the device in order to place the TEC at the desired depth. In the procedure proposed, it will be noted that the degree of complexity of the maneuvers is substantially reduced, as are the means required, which will translate into a substantial reduction in costs. As in the previous section, the installation of the TECs can be divided into the installation sequence of the structure of the TECs and the installation sequence of the gondolas. The following steps have, therefore, been defined in order to perform the automated installation tasks:

- *Installation of the base structure of the TECs:* Note that the structure of the TECs designed for manual maneuvers is the same as the structure designed for automated maneuvers. The installation methodology of the structure of these TECs is, therefore, similar to that used in traditional installation maneuvers.
- *Installation of the gondolas (immersion sequence):* As the gondola is modified to be able to perform automated maneuvers, the installation sequence of the gondolas is different to that used in traditional maneuvers. Figure 9 provides a graphical sequence of the complete process, which is detailed as follows:
  - Once the base structure that supports the gondola is completely installed, the gondola is moved using a tugboat. The tugboat will move one gondola per trip (see Figure 9a).
  - When the tugboat arrives in the position in which the gondola will be placed, the guide cables connected to the structure of the TEC are recovered by using an acoustic signal (see Figure 9b).
  - In the following step, the gondola is lowered onto the seabed. The inner ballast water inside the device is controlled, signifying that the gondola starts filling with water and its inner ballasts permit the gondola to descend. If the difference between the weight of the gondola and the buoyancy force is small, the descent process will be performed slowly. The descent velocity is completely controlled by changing the amount of water inside the inner ballasts (see Figure 9c,d). The gondola descends to a depth close to the base structure that will support the gondola.
  - When the gondola is at the desired depth, but not in a vertical position as regards the base structure owing to the tidal currents, the guide cable will help install the gondola on the base (see Figure 9e).
  - When the gondola is on the structure of the TEC, the inner ballasts of the gondola are filled with water in order to achieve an adequate coupling with the base structure. The installation of the TEC is, therefore, completed (see Figure 9f).

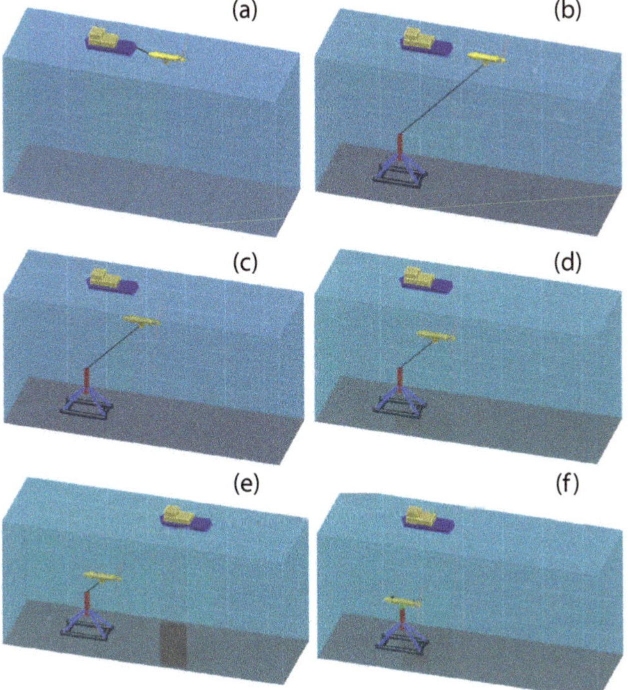

Figure 9. Automated installation maneuver of the gondola (immersion sequence): (**a**) movement of the device with maneuver control in closed loop; (**b**) connection between the cable wire and the gondola; (**c**) immersion maneuver in closed loop; (**d**) immersion maneuver in closed loop (cont.); (**e**) immersion maneuver finished; and (**f**) installation maneuver in closed loop finalized.

The maneuvers that are necessary to perform the maintenance tasks (recovery of a submerged gondola) in a closed loop follow an inverse order to that described for the installation of the gondola. The steps required to perform maintenance maneuvers are the following:

- *Recovery of a submerged gondola (emersion sequence):* Figure 10 shows a visual sequence of the procedure, and a detailed description is provided below.

    - When the gondola is on the structure of the TEC (see Figure 10a), water begins to empty out of the inner ballasts of the gondola until the structure of the TEC and the gondola separate (see Figure 10b).
    - The controlled emersion process of the gondola now begins. Water is emptied out of the inner ballast of the gondola in a controlled manner in order to obtain a smooth emersion movement (see Figure 10c,d). A guide cable is used during the emersion process.
    - When the gondola is on the surface of the sea, the inner ballast is completely emptied in order to increase the buoyancy of the gondola, after which, the control system is disconnected.
    - The maintenance tasks are carried out on the gondola (see Figure 10e). When these tasks have been completed, the control system is connected and the gondola starts filling its inner ballasts with water, thus allowing the gondola to descend automatically (see Figure 10f).
    - All the steps explained for the installation of the gondola (immersion maneuver) are repeated to achieve the connection between the base structure of the TEC and the gondola.

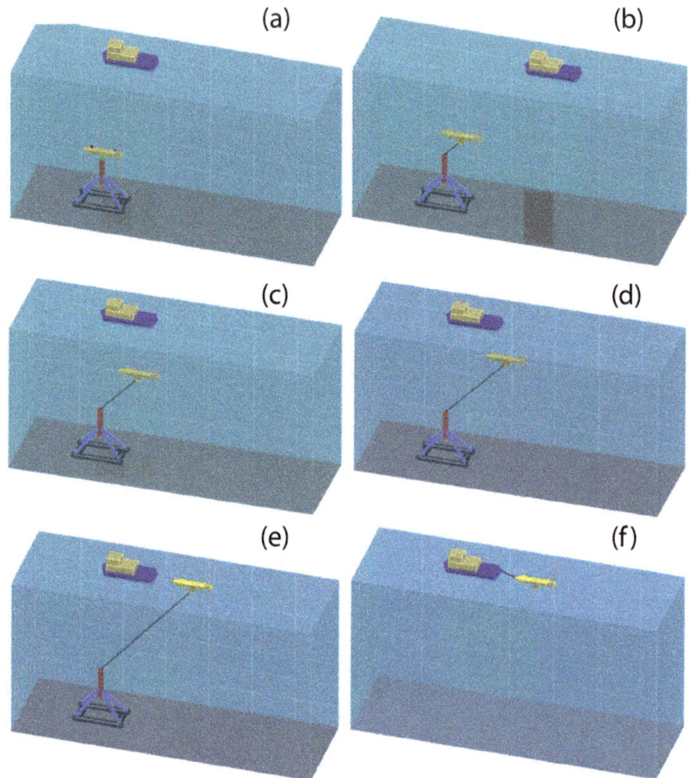

Figure 10. Automated maintenance maneuver of the gondola (emersion sequence): (**a**) the water begins to be emptied out of the gondola; (**b**) separation between the TEC structure and the gondola and start of emersion process; (**c**) emersion maneuver in closed loop; (**d**) emersion maneuver in closed loop (cont.); (**e**) emersion maneuver finalized, the inner ballasts are completely empty and are disconnected from the control system; and (**f**) start of maintenance tasks.

3. Economic-Financial Feasibility Procedure

The objective of this procedure is to forecast the eventual profitability of the investment as regards both the project itself and the stakeholders. This will allow those involved to decide whether or not the project is financially viable. The proposed model, which is shown in Figure 11, has the following stages [19,46]: (i) the LCC of the project is studied and the annual sales estimated; (ii) the financing structure of the model is determined; (iii) the forecast income statement, forecast balance and forecast sources and application of funds for the lifespan of the project are defined; (iv) the cost–benefit analysis is obtained, which is done by utilizing the forecast cash-flows of the project, along with the forecast sources and application funds; (v) the most important financial rations of the model are studied, and (vi) a sensitivity analysis is carried out in order to detect possible business risks. The following subsections deal briefly with this.

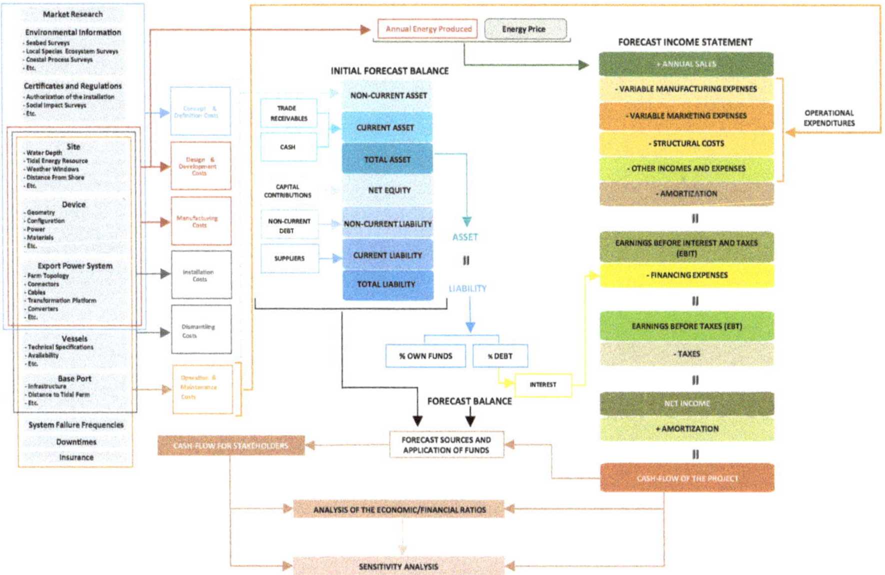

Figure 11. Economic-financial model for gravity-based tidal energy projects.

3.1. Study of the Costs throughout the Service Life of the Project and Estimation of the Annual Sales

The study of the fundamental variables of the economic model implies the definition of the LCC of the project [26,47] and the estimation of the annual sales. Please note that we obtained the estimated costs and estimated annual sales included in the economic feasibility procedure for tidal energy projects (TEPs) by carrying out an analysis of current scientific literature and reports created by companies that specialize in these types of technologies and shipyards.

3.1.1. LCC for TEPs

The LCC is, in essence, an accounting structure containing the mathematical formulations that can be used to estimate the associated costs of the projects during their lifespan [48]. The benefits obtained as a result of determining the LCC of these sorts of projects are summarized as follows [49,50]: (i) the life cycle concept results in earlier actions by which to generate revenue or in lower costs than might otherwise be considered; (ii) better decisions can be made as the result of a more accurate and realistic assessment of revenues and costs, at least within a particular life-cycle stage; (iii) it can promote long-term rewarding in contrast to short-term profitability rewarding; (iv) it provides a better understanding as regards the difference between the acquisition costs and the operating and support costs, and (v) it encourages businesses to find a correct balance between investment costs and operating expenses. In the particular case of TEPs composed of first generation TECs, the stages of which the LCC of the proposed methodology is composed are the following (see also Figure 12). A detailed description of all the structures of the subcosts and the subsequent procedure employed to compute them can be found in [19,46]:

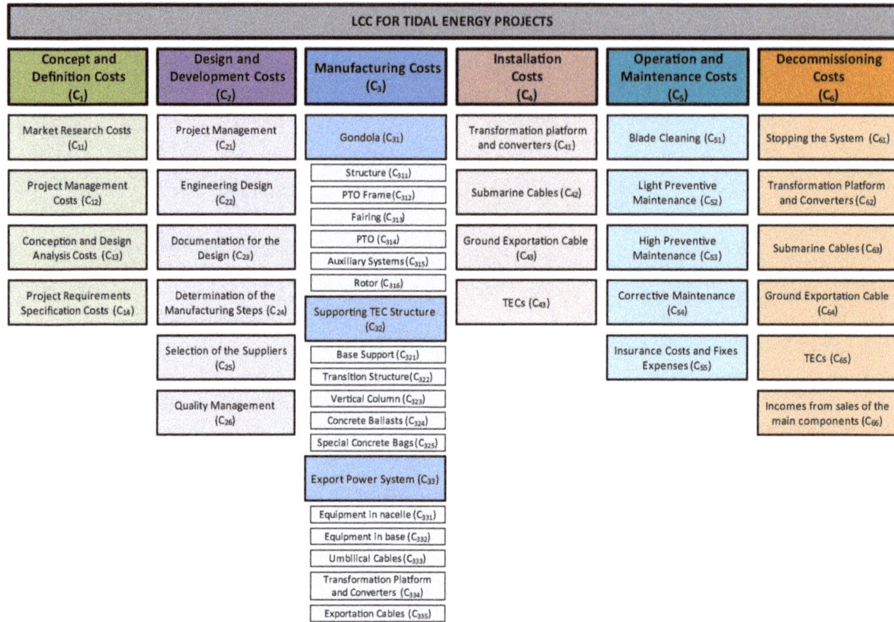

Figure 12. Life cycle costs for tidal energy projects composed of gravity-based first generation TECs.

- *Concept and definition costs* (C_1): These costs correspond to those activities whose objective is to guarantee the feasibility of the project. The costs typically included are the following: (i) market research costs; (ii) project management costs; (iii) conception of the tidal farm and design analysis costs; and (iv) project requirement specification costs.
- *Design and development costs* (C_2): These costs comprise the specification of the requirements of the project and also provide proof that the project has been carried out. They typically include costs regarding: (i) the management of the project; (ii) a technical design and activities for the protection of the environment; (iii) the documentation required for the design; (iv) the definition of the manufacturing steps for the TEF; (v) the inclusion of the selected suppliers; or (vi) quality management.
- *Manufacturing costs* (C_3): These encompass all the costs of manufacturing the elements employed to construct the tidal energy farm and, at this stage, principally consist of: (i) the gondola; (ii) the support structure of the TEC; and (iii) the export power system.
- *Installation costs* (C_4): These are related to the activities required to install all the elements required to construct the TEF, which are, at this stage, typically: (i) the installation of the transformation platform and converters; (ii) the installation of the submarine cables; (iii) the installation of the ground exportation cable; and (iv) the installation of the TECs.
- *Operation and maintenance costs* (C_5): These are all the costs of exploiting the tidal energy project, and include: (i) blade cleaning; (ii) light preventive maintenance; (iii) high preventive maintenance; (iv) corrective maintenance; and (v) insurance costs and fixed expenses.
- *Decommissioning costs* (C_6): These concern the activities required to remove and dispose of the components related to the project, so as to leave the sea as it was before the project started. The principal costs are, in this case: (i) stopping the system; (ii) dismantling the transformation platform and the converters; (iii) dismantling the submarine cables; (iv) dismantling the exportation cable; (v) dismantling the TECs; and (vi) incomes obtained from the sales of the main components (this value will be modeled in the cost structure as a negative value because it is an income).

Once these costs have been computed, the total LCC of a TEP yields the following result:

$$LCC_{TEP} = C_1 + C_2 + C_3 + C_4 + C_5 + C_6. \tag{1}$$

3.1.2. Estimation of the Annual Sales

If the electric tariff is known, then it is possible to estimate the annual sales (AS) attained from a TEP, which are directly related to knowing the annual energy produced (AEP) by the TEF. One of the most important indicators employed to discover whether it will be possible to commercially exploit a particular project is the estimation of the AEP, which usually depends on: (i) the information concerning the site; (ii) the devices used in the TEF; (iii) the energy export system; (iv) the characteristics of the current; (v) the ability of the device to capture energy; and (vi) the ability to convert and export the energy. The procedure employed to accurately estimate the AEP is shown in Figure 13, while the following expressions are required to compute the AEP for a TEF. A detailed description of the procedure developed (based on actuator disk theory) can be found in [46]:

$$
\begin{aligned}
&V_r = V_h \cdot \left(\frac{Z_r}{Z_h}\right)^{\frac{1}{7}}, \\
&a = 1 - \sqrt{1 - c_t}; \quad A_s = \left(\frac{1 - 0.5a}{1 - a}\right) \cdot A_r; \quad V_s = (1 - a) \cdot V_r, \\
&A_s \cdot V_s^2 + (A_t - A_r) \cdot V_r^2 = A_t \cdot V_z^2, \\
&V_x = (V_s - V_z) \cdot e^{-0.2\frac{x}{D}}, \\
&AEP = \sum_{i=1}^{N_{row}} \sum_{j=1}^{N_{column}} P(i,j), \quad \text{where} \quad P(i,j) = \frac{1}{2} \cdot C_P \cdot \rho \cdot A_r \cdot V_x^3(i,j) \cdot \eta_{PTO} \cdot \eta_{AF} \cdot \eta_{PES}, \\
&AS = p_{ET} \cdot AEP,
\end{aligned}
\tag{2}
$$

where V_r symbolizes the velocity of the rotor, V_h expresses the velocity on the sea surface, Z_r denotes the depth of the rotor, Z_h represents the water column depth, a symbolizes the axial induction factor, c_t denotes the thrust coefficient, V_r represents the free stream velocity at the rotor depth, A_r is the surface of the rotor, V_s denotes the flow velocity at the rotor of the turbine, A_s represents the output flow surface, A_t symbolizes the total frontal surface of the TEC, V_z expresses the final velocity of the blended flow, D denotes the diameter of the rotor, (x, V_x) represents an intermediate point located between the rotor output and the blend flow downstream, N_{row} and N_{column} are, respectively, the number of rows and columns on the farm, C_P denotes the power coefficient, $P(i,j)$ denotes the power of the TEC located in row i and column j on the farm, ρ symbolizes the fluid density, η_{PTO} is the performance of the power take off (PTO), η_{AF} denotes the availability factor, η_{PES} represents the performance of the power export system and, finally, p_{ET} is the electric tariff.

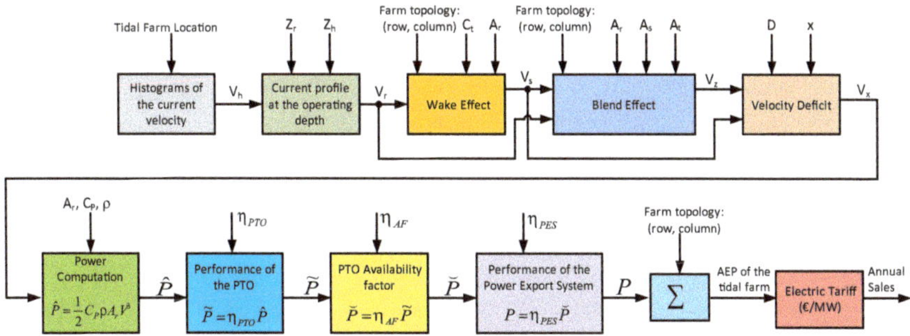

Figure 13. How annual sales are computed: A summary.

3.2. Financing Structure of the Model

The following step involves defining the financial structure employed to define the resources required in the financial model. It is possible to consider the funding sources shown as follows [51]: (i) *own funds*, which will initially consist of the capital invested by stakeholders and, as time progresses, the resources of the project itself, which will consist of the initial stakeholders' contribution and the reserves (project profits which are achieved and not distributed); and (ii) *external resources from financial institutions* that must be returned by the organization.

3.3. Forecast Balance, Forecast Income Statement, and Forecast Sources and Application of Funds in the Model

The way in which the forecast balance, the forecast income statement and the forecast sources and application of funds are obtained is fundamental to understand what will occur, or may occur, in the future of the project [52]. These concepts will be briefly developed below, and a detailed description of these instruments can be found in [46].

3.3.1. Forecast Balance

The tool that allows the clarification of what is occurring in the project on the basis of the available accounting information is the forecast balance, which takes into account the situation of the project's assets, consisting of the project's debts, capital, rights and assets at a particular time. The asset of the forecast balance comprises the rights and the assets, while information regarding the financing attained is obtained from the net equity and the liability. The patrimonial masses of the forecast balance are the following: (i) the *current asset*, which includes those assets that are not initially permanent in the project owing to the fact that their remain period is under a year, and the *non-current asset*, which refers to all those assets that will remain in the project for over a year. The sum of these patrimonial masses, meanwhile, allows us to determine the *total asset* and, therefore, the *total liability*, whose components are the following: (i) the *net equity*, which comprises all the items that are not contemplated as obligations; (ii) the *non-current liability*, which considers all debts that remain in the project for more than a year, and (iii) the *current liability*, which comprises all debts whose remain period is under a year.

3.3.2. Forecast Income Statement

The forecast income statement is employed to quantify an exercise's profit or loss and to discover how it was created. This is done by carrying out an analysis of each item of expenditure with regard to incomes. It is possible to classify the expenses and incomes of this type of projects in the following manner: (i) the *annual sales (AS)*, i.e., the incomes attained after operating the project; (ii) the *operational expenditures (OPEX)*, including operation and maintenance costs, administrative costs, taxes, rent, insurance, etc.; (iii) the *amortization (AM)* during the period as regards plant and equipment, the attrition of property and intangible assets; (iv) the *financing expenses (FE)*, including the project's incomes and financial expenses, and (v) the *corporate taxes (T)*, in which the taxes on the benefits of the period that are different from the other taxes paid by the corporation (and which are normally viewed as structural expenses) are considered. All these expenses are used to organize the forecast income statement in the following manner:

$$
\begin{aligned}
EBIT &= AS - OPEX - AM, \\
EBT &= EBIT - FE, \\
NI &= EBT - T,
\end{aligned}
\tag{3}
$$

where EBT represents earnings before taxes, NI denotes the net income, $EBIT$ symbolizes earnings before interest and taxes and $CF_p = NI + AM$ is the cash-flows of the project.

3.3.3. Forecast Sources and Application of Funds

The forecast sources and application of funds allow us to to attain the variations that took place in the project's patrimonial masses. This is done by contrasting them in two accounting periods with the aim of discovering the sources obtained in that period of time as a consequence of the exploitation cycle of the project, and where they were employed. This makes it possible to discover the project activity's efficiency and comprises: (i) the project resources admitted that year, and how they were applied, and (ii) what effect this movement of inflow and application had on the current asset. The forecast sources and application of funds make it possible to determine the stakeholders' cash-flow, i.e., the cash-flow that it would be possible to disperse in dividends in the case of there being no debt [53].

3.4. Analysis of the Economic and Financial Ratios

By analyzing the economic-financial ratios, it is possible to analyze the project's solvency, liquidity and profitability in order to ascertain its feasibility. We shall, therefore, analyze the ratios shown below so as to determine what strengths and weaknesses these types of projects have [52]:

- *Financial Ratios*: The information obtained in the forecast balance is utilized to determine the project's short-term situation and its liquidity, in addition to its degree of long-term sustainability and its solvency. The following financial ratios will be studied: (i) *Solvency Ratio* (SR), which will make it possible to ascertain how effective the project must be if it is to produce sufficient liquid financial resources in order to punctually meet its commitments as regards the payment of short-term debts resulting from their cycle of operation, in addition to the short-term practicable payments in the same cycle, and (ii) *Total-Debt Ratio (TDR)*, which makes it possible to determine the financial dependence degree by means of the structural composition of funding sources.
- *Economic Ratios*: These make it possible to discover whether the assets are efficiently employed with regard to the management of the operations of said project. The following ratios are employed: (i) *Return of assets (ROA)*, which shows how effective the assets are as regards producing value and; (ii) *Return of Equity (ROE)*, which illustrates how effective the capital contributed by the investors has been. This depends on the net income attained that year.

3.5. Sensitivity Analysis

In order to determine the variables that most affect the viability of the project, it is important for performing a sensitivity analysis with the objective of analyzing the behavior of the project in different situations. The analysis of these scenarios is performed by means of the computation of the *net present value (NPV)*, the *internal rate of return (IRR)* and the *discounted payback period (DPBP)*, which have an extended application in scientific literature [54]. In these sorts of renewable energy projects, the parameters that are, in principle, expected to influence the project profitability are the following: (i) Investment; (ii) AEP; (iii) Price of energy; (iv) Interest rate; (v) Percentage of loan requested, and (vi) Tax rate.

4. Case Study

The installation and maintenance maneuvers defined in Section 2, along with the economic-financial methodology shown in Section 3, will now be applied to a case study consisting of a 50 MW TEF. This farm is composed of gravity-based first generation TECs and is located in the Alderney Race, which is one of the Channel Island Races in the United Kingdom (UK). This will be done in order to determine the viability and profitability of the project from the economic-financial point of view, when using either manual or automated maneuvers, together with the realization of a sensitivity analysis. The analysis will be carried out in an attempt to show that automating these maneuvers will improve the future competitiveness of tidal systems. It may also allow us to detect business risks in the case of the oscillation of fundamental variables of the model, such as investment, the annual energy production, the price of energy, the interest rate, the percentage of loan requested or the tax rate. We shall deal with all of the aforementioned aspects below.

4.1. Description of the Design and Economic-Financial Parameters

The proposed tidal energy farm will consist of 42 1.2 MW TECs, each of which will have an open rotor configuration with their axes parallel to the flow (Figure 14a). It is possible to separate the gondola of the TEC from the structure, which facilitates maintenance tasks. However, the orientation of the gondola is fixed when it is mounted on the structure, and it is consequently mandatory to have a pitch controllable blade system with the aim to maximize the energy that is captured in both current directions. The blades have a diameter of 20 m, while the TEC is attached to the seabed by gravity. The TEF will consist of four rows, and there will be 11, 10, 11, and 10 TECs in each row (Figure 14b). This will make it possible to minimize the shadow effects in the last rows while maximizing the total amount of energy captured by the TEF.

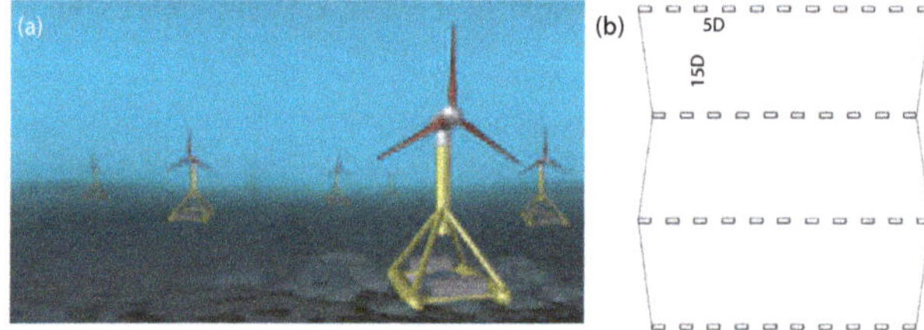

Figure 14. Proposed TEF: (**a**) view and; (**b**) configuration.

The vessel chosen to perform the installation and maintenance procedures on the TEF throughout its service life, which is expected to be 20 years [55], is an HF4 vessel. The base port is that of Cherburg (France), which was chosen for its operative qualities and is located 39 km from the TEF. Finally, the AEP obtained for the TECs in the different rows of the TEF is depicted in Figure 15, in which the following physical parameters have been used to compute an estimation of the AEP of the TEF: $Z_r = 20$ m, $Z_h = 40$ m, $C_t = 0.716$, $A_s = 452$ m^2, $A_r = 314$ m^2, $A_t = 4000$ m^2 (a separation of 100 m between devices and a total depth of 40 m), $D = 20$ m, $x = 30$ m, $C_P = 0.45$, $\rho = 1025$ kg/m^3, $\eta_{PTO} = 0.39$, $\eta_{AF} = 0.97$ and $\eta_{PES} = 0.946$. Furthermore, the case study makes several economic-financial assumptions:

- The electric tariff that has been contemplated is 0.14 €/kWh, increasing by 1.5% every year [16].
- The costs included in the model increase by 1.5% each year.
- The nominal annual discount rate contemplated is 6%, and a value of 2% has also been contemplated for the rate of inflation.
- We assume that: (i) 80% of the investment will be achieved from financing, there will be a fifteen year term and a 3% interest rate for the debt; and (ii) 20% of the total investment will be financed by the partners by means of the project funds.
- The average collection period contemplated is 30 days. The average period of payment is assumed to be 90 days after the service has been provided.
- A system annual depreciation of 5% is assumed in order to achieve a more realistic result.
- The tax rate applied will be 30%.
- There is a particular difficulty as regards the decommissioning costs for TEFs. This is because of the weather windows, the volatility of the costs of the vessels used in this type of operations, the characteristics and uncertainty of offshore operations, etc. Furthermore, there is, at present, no accurate information regarding the quantification of the costs of TEFs because none have been dismantled to date. The decision was, therefore, made not to include the dismantling costs in this case study owing to the aforementioned considerations and uncertainties.

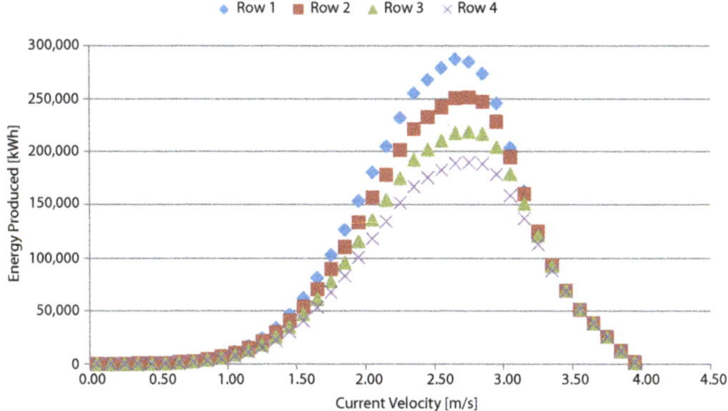

Figure 15. Energy generated by TECs placed in the rows of the TEF.

4.2. Results for Gravity-Based First Generation TECs Maintained with Manual and Automated Maneuvers

Table 1 illustrates the summary of the costs (additional information concerning the achievement of the numerical values of these cost structures can be found in two of the GIT-ERM research group's previous works) [19,46] of the TEF using manual and automated maneuvers. The results concerning the viability and profitability of the proposed project are shown as follows, together with the sensitivity study applied.

Table 1. Summary of the cost of the TEF (Tidal Energy Farm) composed of TECs in the case of manual and automated maneuvers (C_1, C_2, C_3 and C_4 are applied for the first time of the project while C_5 is applied annually).

Manual Maneuvers					
Cost Category					**Total Value (€)**
Concept and Definition Costs (C_1)					**7,350,000**
Design and Development Costs (C_2)					**200,000**
Manufacturing Costs (C_3)					**103,613,936**
gondola					39,563,656
Supporting TEC Structure					21,938,280
Export Power System					42,112,000
Installation Costs (C_4)					**27,700,000**
Transformation Platform and Converters					3,700,000
Submarine and Ground Exportation Cables					7,200,000
TECs					16,800,000
O&M (Operation & Maintenance) Costs (C_5)					**4,905,071**
	Material	Transport	Labour	Production Losses	
Blade Cleaning	0	81,120	4,080	1,256	86,456
Light Preventive Maintenance	142,293	533,513	53,660	32,394	761,860
High Preventive Maintenance	221,784	777,459	39,454	25,669	1,064,366
Corrective Maintenance	0	197,123	7,068	10,919	215,110
Insurance and Fixed Expenses					2,777,279
Decommissioning Costs (C_6)					**0**

Automated Maneuvers					
Cost Category					**Total Value (€)**
Concept and Definition Costs (C_1)					**7,550,000**
Design and Development Costs (C_2)					**300,000**
Manufacturing Costs (C_3)					**105,558,244**
gondola					41,456,364
Supporting TEC Structure					21,938,280
Export Power System					42,163,600
Installation Costs (C_4)					**24,388,000**
Transformation Platform and Converters					3,700,000
Submarine and Ground Exportation Cables					7,200,000
TECs					13,488,000
O&M Costs (C_5)					**4,182,328**
	Material	Transport	Labour	Production Losses	
Blade Cleaning	0	41,371	2,489	1,256	45,116
Light Preventive Maintenance	148,901	277,426	34,342	32,394	493,063
High Preventive Maintenance	231,666	450,926	26,040	25,669	734,301
Corrective Maintenance	0	137,986	5,018	10,918	153,922
Insurance and Fixed Expenses					2,755,926
Decommissioning Costs (C_6)					**0**

4.2.1. Analysis of the Economic-Financial Ratios

Starting by the analysis of the financial ratios, centered on the study of the project liquidity, for which Figure 16 illustrates both the values of the solvency ratio (SR) and the total-debt ratio (TDR) with the use of manual and automated maneuvers during the service life of the project. The comparative results of the SR for the maintenance maneuvers depicted in Figure 16a show that, although the values of these ratios throughout the service life of the project are very high in both cases, the use of automated maneuvers provides higher values than manual maneuvers. We found evidence of a comfortable financial situation for both types of maneuvers, in addition to idle assets that have high opportunity costs. We are, therefore, of the opinion that it would in both cases be possible to invest the idle resources in an effort to lower the opportunity cost and that this would, in turn, allow the project to attain an even higher profitability. With regard to the results obtained for the ROE illustrated in Figure 16b, it will be observed that the results are similar for both maneuvers, with high TDR values in the first years, thus denoting a situation of high indebtedness and, consequently, less protection and greater risks for the creditors. However, said values gradually decrease, and values close to the unit are obtained in the sixth year with a consequent reduction in the probability of insolvency. From this year on, the TDR values are lower than the unit (with null values from the fifteenth year to the end of the project) signifying that the net equity starts to increase substantially. However, although this signifies an excellent solvency, there is, in both cases, a high opportunity cost that could negatively affect the project's profitability, as was explained above. We shall subsequently continue the analysis of the basis of the economic ratios of the project, for which Figure 17 depicts both the return of assets (ROA) and the return of equity (ROE) when using manual and automated maneuvers during the service life of the project. Figure 17a shows that the ROA values for automated maneuvers are higher than in the case of manual maneuvers, and it will be noted that, in both cases, their value is over zero throughout the lifespan of the project. The evolution for automated maneuvers is the same as that for the manual maneuvers, with the values growing successively until the fifteenth year and, from then on, decreasing slightly. After carrying out a detailed analysis of the two subcomponents (economic margin of sales and asset rotation) into which the ROA can be separated for both sorts of maneuvers, we discovered that, during the first fifteen years, the asset rotation grows successively (obtaining higher values for automated maneuvers than for manual maneuvers), but both projects behave in the same way after year fifteen. That is to say, there is a decrease in asset rotation because the growth undergone by the asset is less than that of the sales, signifying that the relationship between both terms decreases with time. Furthermore, the profit from sales increases faster than the economic margin of sales, leading to a growth in the relationship over time during the total service life of the projects. However, from the fifteenth year onward, the economic margin grows less than the decrease in the asset rotation, signifying that the ROA eventually undergoes a slight decrease. It will, nevertheless, always be greater than zero, which is very positive. If we now analyze the results attained for the ROE for manual and automated maneuvers during the service life of the project depicted in Figure 17b, it will be observed that the values of the ROE for automated maneuvers are higher than in the case of manual maneuvers, and this shows that both are positive during the entire service life of the project. However, its value undergoes a slight decrease as time goes by. Upon studying the subcomponents of the ROE (financial sales margin, asset rotation and leverage) in detail, it will be noted that the growth in the asset rotation and the financial sales margin is less than the decrease in financial leverage, and this is the case for both the manual and automated maneuvers. It should, nevertheless, be noted that they all remain positive, signifying that the debt is, in both cases, good for the project. The profitability of the case study for automated maneuvers is thus greater than that obtained in the case of the manual maneuvers. The results also show that investment is recommended for manual and automated maneuvers thanks to the benefits obtained from the first year from both the financial and the accounting points of view.

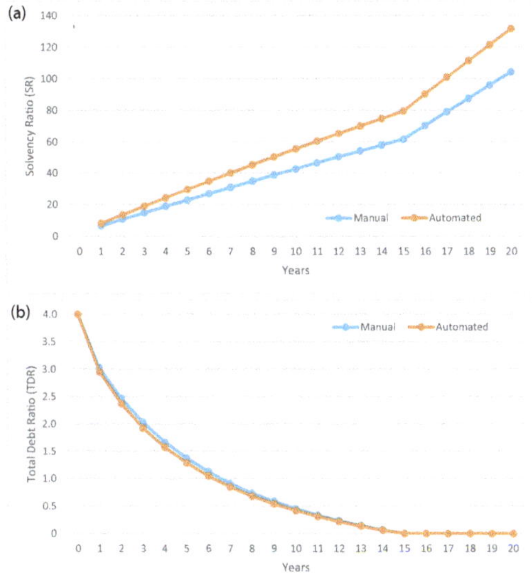

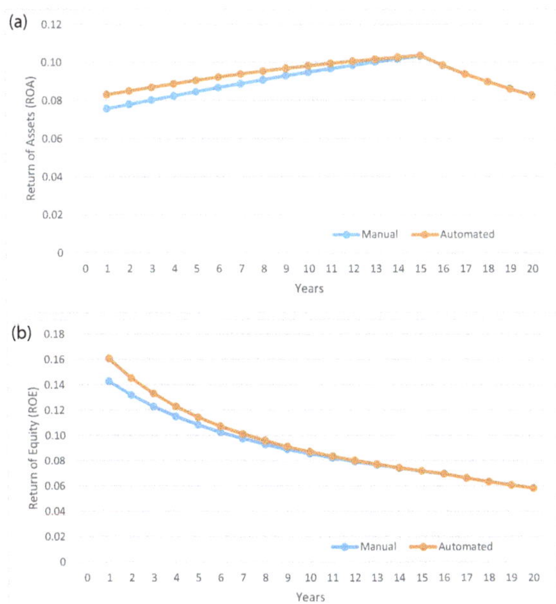

Figure 16. Comparative financial ratios for manual and automated maneuvers: (**a**) solvency ratio (SR), and (**b**) total debt ratio (TDR).

Figure 17. Comparative economic ratios for manual and automate maneuvers: (**a**) return of assets (ROA), and (**b**) return of equity (ROE).

4.2.2. Sensitivity Analysis

After determining those parameters that could have an effect on the profitability of the project (Section 3.5), we shall now identify those that are most critical for the project. The means employed to determine the significance of the aforementioned parameters in the case of the feasibility of the project will be that of altering their values to a slight extent, after which the variations that occur for the reference values considered will be studied. These will, in this case, be the NPV, the IRR and the DPBP of the stakeholders and the project. We shall consider those parameters that lead to higher variations in the references in question to be critical values. In this line of action, the following scenarios have been taken into consideration: (i) an increase in the initial investment to 1%; (ii) a decrease in the AEP by the TEF to 1%; (iii) a decrease in the price of energy to 1%; (iv) an increase in the interest rate to 1%; (v) a decrease in the percentage of the loan requested to 1%; and (vi) an increase in the tax rate to 1%. The values obtained for the nominal case for manual values are the following: the NPV for the project is 64,066,094 € and the NPV for the stakeholders is 94,399,360 €; the IRR for the project is 8.126% and the IRR for the stakeholders is 24.17%, and the DPBP for the project is 13.04 years and the DPBP for the stakeholders is 4.95 years. The values obtained for the nominal case for automated maneuvers are, meanwhile, the following: the NPV for the project is 75,432,173 € and the NPV for the stakeholders is 105,494,103 €; the IRR for the project is 8.21% and the IRR for the stakeholders is 24.17%, and the DPBP for the project is 12.23 years and the DPBP for the stakeholders is 4.37 years. Figures 18 and 19 depict the results of the NPV sensitivity analysis with regard to the project and the stakeholders when manual and automated maneuvers are carried out, while Table 2 shows the variations in the NPV, IRR and DPBP of the project and the stakeholders as regards manual and automated maneuvers in the aforementioned scenarios.

These results allow us to conclude that, in the nominal case, the TEP carried out using manual and automated maneuvers are economically feasible, with a higher profitability for automated maneuvers. Furthermore, both the automated and the manual maneuvers are affected by the same parameters as those that affect the profitability of the project, which are the variations in interest rate, the price of energy and the AEP. This is owing to the fact that these parameters concern the financing of the farm, and the incomes will be achieved during the entire service life of the project. The profitability of the project is, however, less affected by the variations related to the tax rate, the percentage of the loan requested and the investment.

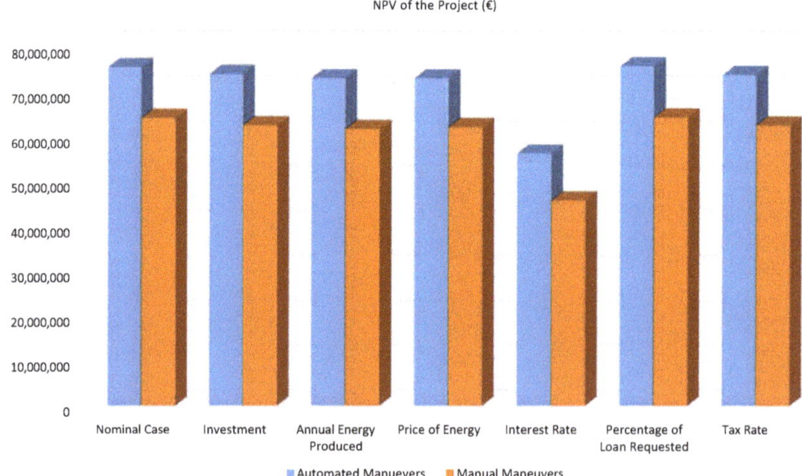

Figure 18. NPV (Net Present Value) results of the project when manual and automated maneuvers are carried out.

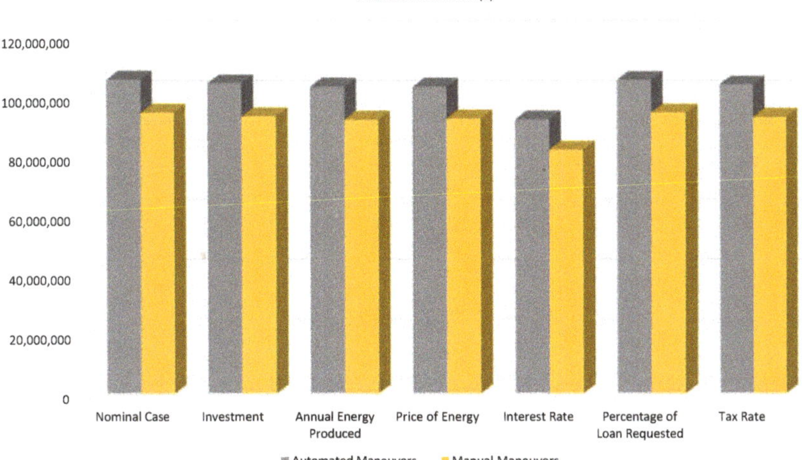

Figure 19. NPV results for the stakeholders when manual and automated maneuvers are carried out.

Table 2. Sensitivity analysis results when manual and automated maneuvers are carried out.

	Manual Maneuvers					
	Project			**Stakeholders**		
	Δ_{NPV}	Δ_{IRR}	Δ_{DPBP}	Δ_{NPV}	Δ_{IRR}	Δ_{DPBP}
Investment is increased to 1%	−2.46%	−1.66%	1.23%	−1.29%	−2.0%	2.49%
Annual energy produced by the farm decreases to 1%	−3.82%	−1.84%	1.38%	−2.52%	−2.23%	2.79%
Price of the energy decreases to 1%	−3.14%	−1.52%	1.13%	−2.07%	−1.83%	2.28%
Interest rate increases to 1%	−28.56%	∼0%	7.88%	−13.21%	∼0%	3.09%
Percentage of the loan requested decreases to 1%	−0.34%	−0.20%	0.17%	−0.17%	−1.40%	3.61%
Tax rate increases to 1%	−2.41%	−0.20%	0.78%	−1.57%	−1.40%	1.35%
	Automated Maneuvers					
	Project			**Stakeholders**		
	Δ_{NPV}	Δ_{IRR}	Δ_{DPBP}	Δ_{NPV}	Δ_{IRR}	Δ_{DPBP}
Investment is increased to 1%	−2.06%	−1.64%	1.22%	−1.18%	−2.0%	2.33%
Annual energy produced by the farm decreases to 1%	−3.28%	−1.84%	1.32%	−2.34%	−2.23%	1.69%
Price of the energy decreases to 1%	−3.27%	−1.52%	1.32%	−2.34%	−1.83%	2.49%
Interest rate increases to 1%	−25.45%	∼0%	7.36%	−12.76%	∼0%	2.77%
Percentage of the loan requested decreases to 1%	−0.29%	−0.18%	0.16%	−0.15%	−1.41%	3.6%
Tax rate increases to 1%	−2.25%	−0.18%	0.82%	−1.61%	−1.41%	1.35%

4.3. Comparative Sensitivity Analysis

Having employed a case study to demonstrate the excellent profitability and economic feasibility as regards the use of both manual and automated maneuvers, we shall now develop a comparative sensitivity analysis in order to quantify, in an economic manner, how much better the use of automated maneuvers is with regard to the use of manual maneuvers. As occurred in Section 4.2.2, we shall study the following scenarios: (i) the nominal case; (ii) an increase in the initial investment to 1%; (iii) a decrease in the AEP by the TEF to 1%; (iv) a decrease in the price of energy to 1%; (v) an increase in the interest rate to 1%; (vi) a decrease in the percentage of the loan requested to 1%; and (vii) an increase in the tax rate to 1%. Table 3 shows the results of the comparative sensitivity analysis carried out for the variations in the NPV, IRR and DPBP of the project and the stakeholders in the case of the use of automated maneuvers when compared to the use of manual maneuvers in the aforementioned scenarios. These results allow us to conclude that the profitability of the case study is higher and the investment is recovered faster when using automated maneuvers for all scenarios, demonstrating that, from an economic point of view, this is an attractive solution as regards a future commercialization of these devices.

Table 3. Comparative sensitivity analysis results for the use of automated maneuvers when compared to the use of manual maneuvers.

	Project			Stakeholders		
	Δ_{NPV}	Δ_{IRR}	Δ_{DPBP}	Δ_{NPV}	Δ_{IRR}	Δ_{DPBP}
Nominal Case	17.74%	∼0%	−6.23%	11.75%	∼0%	−11.45%
Investment is increased to 1%	18.24%	∼0%	−6.23%	11.94%	∼0%	−11.74%
Annual energy produced by the farm decreases to 1%	18.39%	∼0%	−6.25%	12.03%	∼0%	−11.84%
Price of the energy decreases to 1%	17.58%	∼0%	−6.02%	11.52%	∼0%	−11.41%
Interest rate increases to 1%	22.88%	∼0%	−6.64%	12.40%	∼0%	−11.86%
Percentage of the loan requested decreases to 1%	18.48%	∼0%	−6.51%	11.77%	∼0%	−11.46%
Tax rate increases to 1%	17.93%	∼0%	−6.16%	11.78%	∼0%	−11.60%

5. Conclusions and Future Works

The development of improved installation and maintenance procedures for technologies that harness energy from ocean currents is a promising research field that needs to be studied in detail in order to achieve a successful future commercialization of these technologies. In this paper, we have explained the procedures for manual and automated installation and maintenance maneuvers for gravity-based first generation TECs, and have also carried out an economic-financial evaluation in order to highlight the merits of automated installation and maintenance maneuvers for tidal energy technologies. Some of the potential benefits that the use of these maneuvers could provide are the following: (i) an increase in the competitiveness of these technologies; (ii) a reduction in human resources and the number and duration of installation and maintenance operations; (iii) the use of cheaper general purpose ships instead of high cost special vessels for maintenance operations; and (iv) a higher project profitability, among others. We have conducted a numerical case study of a TEF in the Alderney Race (UK). This farm comprises 42 TECs of 1.2 MW, and the study was carried out with the objective of determining the economic-financial viability of the project when employing either automated or manual installation and maintenance maneuvers. After applying the economic-financial procedure explained throughout this paper to the case study, we attained results indicating that the project would, in both cases, obtain a good profitability and can consequently recommend investment from both accounting and financial points of view, since the benefits produced from the first year are good. We have additionally discovered that the profitability is even greater when employing devices capable of performing automated maneuvers. The aforementioned results also indicate that the variables that have the greatest influence on the profitability of the project are, in the case of both types of maneuvers, the AEP, the price of energy and the interest rate. Finally, having attained the results of the case study, we can conclude that the economic-financial methodology employed is useful as regards modeling different renewable energy projects, as is also the case of the manual and automated installation and maintenance procedures for first generation TECs presented in this work. This methodology additionally has an advantage in that it could, with minor adaptations, be used for other kinds of offshore renewable energy projects and TEC designs. Our future research will be focused on the application of this model to other offshore renewable energy projects.

Author Contributions: E.S., R.M. and J.A.S. conceived, designed and performed the proposed methodology and the case study. Additionally, E.S., R.M. and J.A.S. analyzed the data and participated in writing the paper.

Funding: This research has been supported by the Spanish Ministerio de Economía y Competitividad under Research Grants DPI2014-53499-R.

Conflicts of Interest: The authors declare no conflicts of interest.

Abbreviations

The following abbreviations are used in this manuscript:

AEP	Annual Energy Produced
AM	Amortization
AS	Annual Sales
DPBP	Discounted Payback Period
EBT	Earnings Before Taxes
EU	European Union
FE	Financing Expenses
GIT-ERM	Grupo de Investigación Tecnológico en Energías Renovables Marinas
IRR	Internal Rate of Return
LCC	Life Cycle Costs
OPEX	Operational Expenditures
MRE	Marine Renewable Energy
NI	Net Income
NPV	Net Present Value
O & M	Operation and Maintenance
ROA	Return of Assets
ROE	Return of Equity
ROV	Remotely Operated Vehicle
SR	Solvency Ratio
T	Corporate Taxes
TDR	Total-Debt Ratio
TEC	Tidal Energy Converter
TEF	Tidal Energy Farm
TEP	Tidal Energy Projects
UK	United Kingdom

References

1. Rodríguez-Delgado, C.; Bergillos, R.A.; Iglesias, G. Dual wave farms for energy production and coastal protection under sea level rise. *J. Clean. Prod.* **2019**, *222*, 364–372. [CrossRef]
2. Atilgan, B.; Azapagic, A. Life cycle environmental impacts of electricity from fossil fuels in Turkey. *J. Clean. Prod.* **2015**, *106*, 555–564. [CrossRef]
3. Sequeira, T.N.; Santos, M.S. Renewable energy and politics: A sistematic review and new evidence. *J. Clean. Prod.* **2018**, *192*, 553–568. [CrossRef]
4. Sinha, A.; Shahbaz, M.; Sengupta, T. Renewable energy and policies and contratictions in causality: A case of next 11 countries. *J. Clean. Prod.* **2018**, *197*, 73–84. [CrossRef]
5. Directive 2009/28/EC of the European Parliament and of the Council of 23 April 2009 on the promotion of the use of energy from renewable sources and amending and subsequently repealing directives 2001/77/EC and 2003/30/EC. *Off. J. Eur. Union* **2009**, 16–60.
6. Magagna, D.; MacGillivray, A.; Jeffrey, H.; Hanmer, C.; Raventos, A.; Badcock-Broe, A.; Tzimas, E. *Wave and Tidal Energy Strategic Technology Agenda*; SI Ocean: Brussels, Belgium, 2014.
7. Jeffrey, H.; Jay, B.; Winskel, M. Accelerating the development of marine energy: Exploring the prospects, benefits and challenges. *Technol. For. Soc. Chang.* **2013**, *80*, 1306–1316. [CrossRef]
8. *United Nations Framework Convention on Climate Change, 2015. Paris Agreement*; UNFCCC Secretariat: Bonn, Germany, 2016.
9. *Overcoming Research Challenges for Ocean Renewable Energy*; Energy Research Knowledge Centre: Brussels, Belgium, 2013.
10. Ocean Energy Strategic Roadmap 2016, Building Ocean Energy for Europe. *Ocean Energy Forum*, 8 November 2016.
11. Hardisty, J. *The Analysis of Tidal Stream Power*; Wiley: Hoboken, NJ, USA, 2009; ISBN 978-0-470-72451-4.

12. Portilla, M.P.; Somolinos, J.A.; López, A.; Morales, R. Modelado dinámico y control de un dispositivo sumergido provisto de actuadores hidrostáticos. *Revista Iberoamericana de Automática e Informática Industrial* **2018**, *15*, 12–23. [CrossRef]

13. Alstom Tidal Turbines Web Page. Available online: https://marineenergy.biz/tag/alstom/ (accessed on 3 April 2019).

14. Andritz Hydro Hammerfest. How It Works. Available online: http://www.andritz.com/hy-hammerfest.pdf (accessed on 3 April 2019).

15. Fallon, D.; Hartnett, M.; Olbert, A.; Nash, S. The effects of array configuration on the hydro-environmental impacts on tidal turbines. *Renew. Energy* **2014**, *64*, 10–25. [CrossRef]

16. Segura, E.; Morales, R.; Somolinos, J.A. A strategic analysis of tidal current energy conversion systems in the European Union. *Appl. Energy* **2018**, *212*, 527–551. [CrossRef]

17. Denny, E. The economics of tidal energy. *Energy Policy* **2009**, *37*, 1914–1924. [CrossRef]

18. Segura, E.; Morales, R.; Somolinos, J.A.; López, A. Techno-economic challenges of tidal energy conversion systems: Current status and trends. *Renew. Sustain. Energy Rev.* **2017**, *77*, 536–550. [CrossRef]

19. Segura, E.; Morales, R.; Somolinos, J.A. Cost assessment methodology and economic viability of tidal energy projects. *Energies* **2017**, *10*, 1806. [CrossRef]

20. Nautricity Web Page, 2016. Available online: http://www.nautricity.com/cormat/ (accessed on 3 April 2019).

21. Tocardo Web Page, 2016. Available online: http://www.tocardo.com/ (accessed on 3 April 2019).

22. Somolinos, J.A.; López, A.; Portilla, M.P.; Morales, R. Dynamic model and control of a new underwater three-degree-of-freedom tidal energy converter. *Math. Probl. Eng.* **2015**, *2015*, 948048. [CrossRef]

23. López, A.; Somolinos, J.A.; Núñez, L.R.; Morales, R. Dynamic Model and Experimental Validation for the Control of Emersion Maneuvers of Devices for Marine Currents Harnessing. *Renew. Energy* **2017**, *103*, 333–345.

24. Morales, R.; Fernández, L.; Segura, E.; Somolinos, J.A. Maintenance Maneuver Automation for an Adapted Cylindrical Shape TEC. *Energies* **2016**, *9*, 746. [CrossRef]

25. Fernández, L.; Segura, E.; Portilla, M.P.; Morales, R.; Somolinos, J.A. Dynamic model and nonlinear control for a two degrees of freedom first generation tidal energy converter. *IFAC-PapersOnLine* **2016**, *49–23*, 373–379. [CrossRef]

26. Castro-Santos, L.; Filgueira-Vizoso, A.; Lamas-Galdo, I.; Carral-Couce, L. Methodology to calculate the installation costs of offshore wind farms located in deep waters. *J. Clean. Prod.* **2018**, *170*, 1124–1135. [CrossRef]

27. Castro-Santos, L.; Silva D.; Rute Bento, A.; Salvaçao, N.; Guetes Soares, C. Economic Feasibility of Wave Energy Farms in Portugal. *Energies* **2018**, *11*, 3149. [CrossRef]

28. Castro-Santos, L.; Martins, E.; Guedes-Soares, C. Cost assessment methodology for combined wind and wave floating offshore renewable energy systems. *Renew. Energy* **2016**, *97*, 866–880. [CrossRef]

29. Castro-Santos, L.; Martins, E.; Guedes-Soares, C. Economic comparison of technological alternatives to harness offshore wind and wave energies. *Energy* **2017**, *140*, 1121–1130. [CrossRef]

30. Voith. Tidal Current Power Stations. Available online: http://voith.com/en/productsservices/hydro-power/ocean-energies/tidal-current-power-stations--591.html (accessed on 3 April 2019).

31. Tidal Energy: Technology Brief. *International Renewable Energy Agency (IRENA)*, June 2014. Available online: https://www.irena.org/documentdownloads/publications/tidal_energy_v4_web.pdf (accessed on 4 April 2019).

32. BVG Associates. A Guide to an Offshore Wind Farm. *The Crown Estate*, 2010. Available online: http://www.thecrownestate.co.uk/media/5408/ei-a-guide-to-an-offshore-wind-farm.pdf (accessed on 4 April 2019).

33. TradeWinds. The Global Shipping News source, 2017. Available online: http://www.tradewindsnews.com/ (accessed on 4 April 2019).

34. Altlantis Resources. AR1000, 2017. Available online: https://www.atlantisresourcesltd.com/services/turbines/ (accessed on 4 April 2019).

35. ABR Company Ltd. International Tug & OSV, Incorporating Salvage News, 2017. Available online: https://www.tugandosv.com/about_the_magazine.php (accessed on 4 April 2019).

36. Somolinos, J.A. Control de operaciones de dispositivos marinos de aprovechamiento de la energía hidrocinética. Proyecto RETOS de la Sociedad DPI2014m bn-53499-R. 2015. Available online: http://www.upm.es/observatorio/vi/index.jsp?pageac=grupo.jsp&idGrupo=391 (accessed on 20 May 2019).
37. Espín, M. Modelado Dinámico y Control de Maniobras de Dispositivos Submarinos. Ph.D. Thesis, ETSIN-UPM, Madrid, Spain, 2015.
38. Sánchez, G. Diseño de un dispositivo para el aprovechamiento de la energía de las corrientes (DAEC) y su integración en un parque marino. Master's Thesis, Escuela Técnica Superior de Ingenieros Navales, Universidad Politécnica de Madrid (ETSIN-UPM), Madrid, Spain, 2014.
39. Subsea Power Cables in Shallow Water Renewable Energy Applications. *Det Norske Veritas (DNV) AS*, February 2014. Available online: https://rules.dnvgl.com/docs/pdf/DNV/codes/docs/2014-02/RP-J301.pdf (accessed on 29 May 2019).
40. Bryden, I.G.; Couch, S.J. ME1–Marine energy extraction: Tidal resource analysis. *Renew. Energy* **2006**, *31*, 133–139. [CrossRef]
41. Charlier, R.H. A Sleeper awakes: Tidal current power. *Renew. Sustain. Energy Rev.* **2003**, *7*, 515–529. [CrossRef]
42. Marine Current Turbines (MCT). An Atlantis Company. *Tidal Energy Section*, 2017. Available online: http://www.marineturbines.com/Tidal-Energy (accessed on 4 April 2019).
43. López, A.; Somolinos, J.A.; Núñez, L.R. Dispositivo para el aprovechamiento de las corrientes marinas multi-rotor con estructura poligonal Patent Number P201430182. ES2461440, 25 November 2014.
44. López, A.; Somolinos, J.A.; Núñez, L.R. Underwater electrical generator for the harnessing of bidirectional flood currents. U.S. Patent Application 12/978993, 30 June 2011.
45. López, A.; Núñez, L.R.; Somolinos, J.A. Generador electrico submarino para el aprovechamiento de las corrientes de flujo bidireccional. Patent Number ES 2341311, 30 December 2009.
46. Segura, E.; Morales, R.; Somolinos, J.A. Economic-Financial Modeling for Marine Current Harnessing Projects. *Energy* **2018**, *158*, 859–880. [CrossRef]
47. Delogu, M.; Zanchi, L.; Maltese, S.; Boloni, A.; Pierini, M. Environmental and economic life cycle assessment of a lightweight solution for an automotive component: A comparison between talc-filled and hollow glass microspheres-reinforced polymer composites. *J. Clean. Prod.* **2016**, *139*, 548–560. [CrossRef]
48. *IEC 60300-3-3:2004 - Dependability Management - Part 3-3: Application Guide - Life Cycle Costing*; IEC: Geneva, Switzerland, 2004.
49. Dhillon, B.S. *Life-Cycle Costing for Engineers*; CRC Press: Boca Raton, FL, USA, 2010.
50. Vail Farr, J. *Systems Life Cycle Costing—Economic Analysis, Estimation and Management*; CRC Press: Boca Raton, FL, USA, 2011.
51. Hirt, G.; Block, S. *Fundamentals of Investment Management*, 10th ed.; McGraw-Hill Education: New York, NY, USA, 2011.
52. Kimmel, P.D.; Weygandt, J.J.; Kieso, D.E. *Accounting: Tools for Business Decision Making*, 4th ed.; John Wiley & Sons Ltd: Hoboken, NJ, USA, 2011.
53. Wild, J.J. *Financial Accounting Fundamentals*, 6th ed.; McGraw-Hill Education: New York, NY, USA, 2017.
54. Short, W.; Packey, D.; Holt, T. *A Manual for the Economic Evaluation of Energy Efficiency and Renewable Energy Technologies*; NREL/TP-462-5173; National Renewable Energy Laboratory: Colorado, CO, USA, 1995.
55. Det Norske Veritas AS. Det Norske Veritas (DNV), DNV–OS–J101. *Design of Offshore Wind Turbine Structures*; Det Norske Veritas AS.: Oslo, Norway, 2010; pp. 1–142.

 © 2019 by the authors. Licensee MDPI, Basel, Switzerland. This article is an open access article distributed under the terms and conditions of the Creative Commons Attribution (CC BY) license (http://creativecommons.org/licenses/by/4.0/).

 *energies*

Article

A Discussion on the Effective Ventilation Distance in Dead-End Tunnels

Manuel García-Díaz [1], Carlos Sierra [2,*], Celia Miguel-González [1] and Bruno Pereiras [1]

[1] Department of Energy, Polytechnic School of Engineering of Gijón, University of Oviedo, 33203 Gijón, Spain
[2] Department of Mining, Topography and Structure Technology, University of León, 24071 León, Spain
* Correspondence: csief@unileon.es

Received: 16 July 2019; Accepted: 27 August 2019; Published: 30 August 2019

Abstract: Forcing ventilation is the most widely used system to remove noxious gases from a working face during tunnel construction. This system creates a region near the face (dead zone), in which ventilation takes place by natural diffusion, rather than being directly swept by the air current. Despite the extensive use of this system, there is still a lack of parametrical studies discerning the main parameters affecting its formation as well as a correlation indicating their interrelation. With this aim in mind, computational fluid dynamics (CFDs) models were used to define the dead zone based on the airflow field patterns. The formation of counter vortices, which although maintain the movement of air hinder its renewal, allowed us to discuss the old paradigm of defining the dead zone as a very low air velocity zone. Moreover, further simulations using a model of air mixed with NO_2 offered an idea of NO_2 concentrations over time and distance to the face, allowing us to derive at a more realistic equation for the effective distance. The results given here confirm the degree of conservativism of present-day regulations and may assist engineers to improve ventilation efficiency in tunnels by modifying the duct end-to-face distance.

Keywords: ancillary ventilation; effective zone; CFDs; mixture model

---

1. Introduction

Underground ventilation provides enough airflow to the workings to dilute and remove noxious gases and dust. Furthermore, underground ventilation supplies O_2 where it has been depleted and controls temperature. Noxious gases may come from strata rock (e.g., CH_4, CO_2, H_2S, Rn), be generated by underground machinery (e.g., CO, CO_2, NO_x), or be a consequence of blasting operations (CO, NO_x) [1,2]. Two systems are used for mine ventilation purposes, namely, principal and ancillary. The former provides air to the overall mine, establishing general air circulation. The latter is used in developing workings and galleries and provides fresh air to specific areas in the mine. The ancillary ventilation of mine galleries is similar to that used in tunnels under construction. The main difference between them is that, for tunnel ventilation purposes, outside air is directly conducted to the face, whereas in mine ventilation fresh air has to be diverted from the main ventilation system into the working faces.

The favorite ventilation system for ancillary mine ventilation is the forcing system. Forcing is preferred over exhaust ventilation, as the fresh air coming along the roadway usually enters the duct straight away without sweeping the face first [3] (p. 7). The physics of this ventilating system is similar to a that of free jet that progressively opens, reducing its velocity until it reaches zero. The region in which the air is immobile is termed "the dead zone", and it is separated by distance from the face "d_z" (Figure 1). In this zone, ventilation takes place by natural diffusion rather than being directly swept by the air current. The distance between the duct end and the dead zone is the air jet range, which is termed "the effective distance" (e_z). The sum of both distances is the space between the duct end and the face (d_s). The limit case occurs when there is no dead zone so that "e_z" is equal to "d_s".

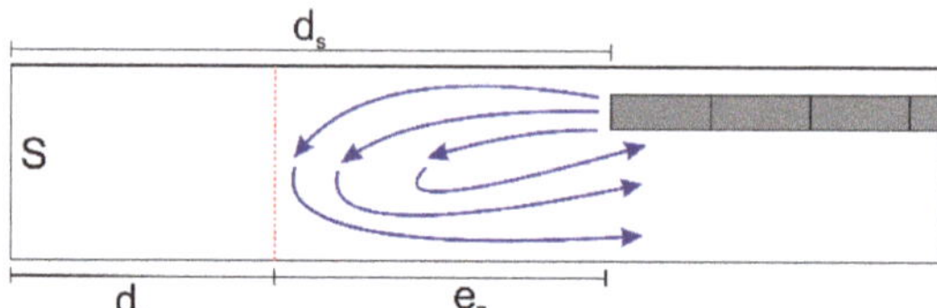

Figure 1. Section (S), dead (d_z) and effective zones (e_z) for a drivage forcing ventilation system.

The most simple reference range for e_z is less than 15 m (e.g., [3] (p. 7)). Other authors consider "e_z" as a function of the gallery section (S) so that e_z be less than $4\sqrt{S}$ or less than $6\sqrt{S}$ (e.g., [4] (p. 381)). Moreover, others consider the gallery section and a parameter "a", which is usually 0.07, in which case $e_z < 0.5\sqrt{S\left(1+\frac{1}{2a}\right)} = 4\sqrt{S} \simeq 12-15$ m [5] (p. 245). Finally, additional approaches consider the ventilation duct as a reference, in which case "e_z" should be 10–15 times as large as its diameter [6]. All legal requirements worldwide follow one of these recommendations, without getting into the physics of the dead zone formation.

Computational fluid dynamics (CFDs) has become an important tool for most fluid dynamics engineering problems, thus playing an emerging role in mine ventilation systems design [7]. Its popularity is mainly due to its low cost and capability of measuring parameters, which are difficult or almost impossible to obtain experimentally, as well as its fast assessment and adaptability to varying design conditions. In this context, CFDs may offer insight into the mechanics of dead zone formation.

The first CFDs in the field of mine ventilation were started by Herdeen and Sullivan (1993) and Srinivasa et al. (1993), who used a Eulerian–Lagrangian formulation. Since the apparition of modern commercial software, CFDs models have been used in several aspects of mine ventilation. Thus, extensive studies have been conducted in the field of dust transport (e.g., [7,8]) and heat transfer. The latter is important the case of heat fires and explosions in longwall working faces (e.g., [9,10]). Moreover, Sasmito et al. [11] studied air conditioning as a complement to ventilation for heat control in dead-end tunnels. Furthermore, Yuan and Smith [12] studied the low-temperature oxidation of coal using unsteady-state simulations. Likewise, Shi et al. [13] developed a CFDs model for coal spontaneous combustion under goaf gas drainage conditions.

As concerns auxiliary ventilation, Diego et al. [14] compared traditional and CFDs models for calculating losses in a dead-end circular tunnel, highlighting the advantages of CFDs over traditional calculations. Toraño et al. [15] studied auxiliary ventilation roadways driven with roadheaders, but their results cannot be directly transferred for dead-end tunnel ventilation. Sasmito et al. [16] performed a computational study on gas control in a room and pillar coal mine. Toraño et al. (2009) [17] modeled methane behavior, but did not consider blasting gases, in coal mines ancillary ventilation. Li, Aminossadati and Wu [18] studied ancillary ventilation in super large developments. Onder, Sarac and Cevik studied the ventilation of a cul-de-sac, but focused on fan–duct interaction rather than on the flow outside the duct. Fang, Yao and Lei conducted a parametrical study but did not consider the cross-sectional area as a variable [19]; thus, the magnitude of the effects associated with the whole set of parameters is yet to be established. Szlazak et al. [20] and Reed and Taylor [21] indicated that ancillary ventilation has been conducted under old-fashioned guidelines. As a consequence, it can be concluded that there is still a lack of knowledge concerning ventilation setup influence on dead zone formation in an empty heading. All things considered, the main aim of this study is to determine the correlations among the main factors influencing the effective distance in cul-de-sac ventilation so that their respective influences can be quantified. The following parameters are taken into account: tunnel section, flow rate and position with regard to the tunnel axis. This distance was first assessed based on the flow field patterns obtained, and later on a mixture model of air and NO_2.

The rest of the paper is organized as follows. Section 2 defines the geometries, exposes the discretization, presents the equations solved by the CFDs model, defines the boundary conditions, indicates the software configuration, describes the criteria followed for the delimitation of the dead zone and explains the mixture model. Section 3 discusses the delimitation of the dead zone and presents the correlation for dead-end-to-face distance, based on both flow field patterns and flow field patterns plus the mixture model correction. We conclude the paper in Section 4.

2. Materials and Methods

All the work was carried out by means of the ANSYS package, which allows for the modeling of different engineering problems using numerical methods. In this case, the hardware used to run the bulk of the simulations consisted in a cluster of computers equipped with i7-6800K processors and 32 GB DDR4 RAM. With these computers and the settings described later, each simulation was run for between 4 h and 5 h. Figure 2 illustrates a flow chart depicting the methodology followed by the authors.

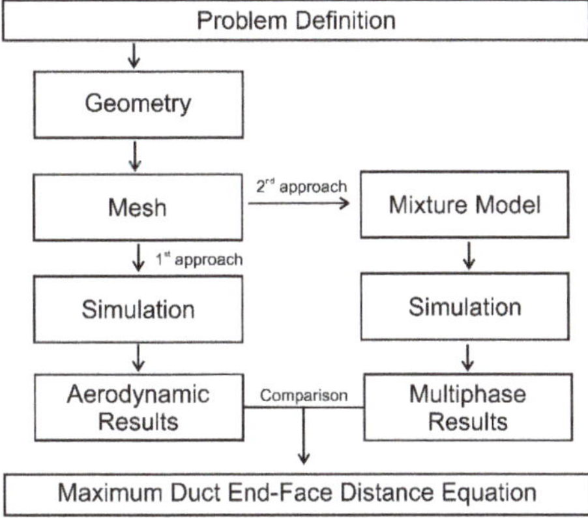

Figure 2. Flow chart that depicts the methodology followed.

2.1. Geometries

The geometry was generated using an ANSYS Design Modeler, commercial software, which is included in the ANSYS WorkBench package. A parametric study was performed for the study. The geometries of the tunnel and the variables considered are given in Figure 3. Face-to-duct end distances were large enough to ensure that the flow was not influenced by boundary conditions. The tunnel profile was defined for a regular tunnel section. All of the analyzed profiles were concentric to that depicted in Figure 3. Common sections for mine ducts were retrieved from commercial catalogs and particularly from [22] (p. 236). The parametrized geometry is specified in Table 1 for the four scenarios considered.

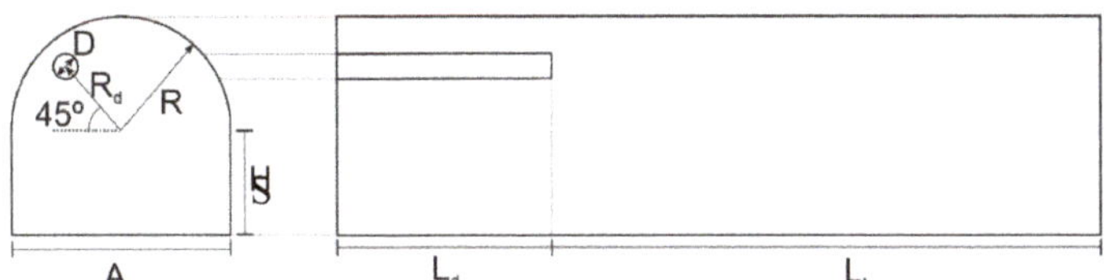

Figure 3. Definition of the parameters. Tunnel view: Left, lateral; right, front.

Table 1. Geometrical parameters used for the parametric study.

Geometrical Parameters for Each Scenario	Parameter							
	A	**H**	**R**	**R$_d$**	**D**	**L$_d$**	**L$_t$**	**S**
Units					m			m^2
	1	5.4	1.8	2.7	1.8	0.54			21.17
Scenario	2	6	2	3	2	0.60	50	100	26.13
	3	7.2	2.4	3.6	2.4	0.72			37.63
	4	9	3	4.5	3	0.90			58.80

2.2. Discretization

The geometry discretization was performed by means of ANSYS ICEM CFD. This software allows for the formation of high-quality hexahedral meshes based on the blocking method. Each of the meshes generated for the four scenarios had about one million cells (Figure 4a). The mesh quality was assessed by means of the angle criterion. This criterion takes into account the internal angles of each cell and seeks the maximum internal angle deviation from 90° for each element. Figure 4b shows that the mesh quality was above the typical quality criterion for CFDs purposes, which is 18°. In addition, most of the cell angles were close to the optimum value (90°).

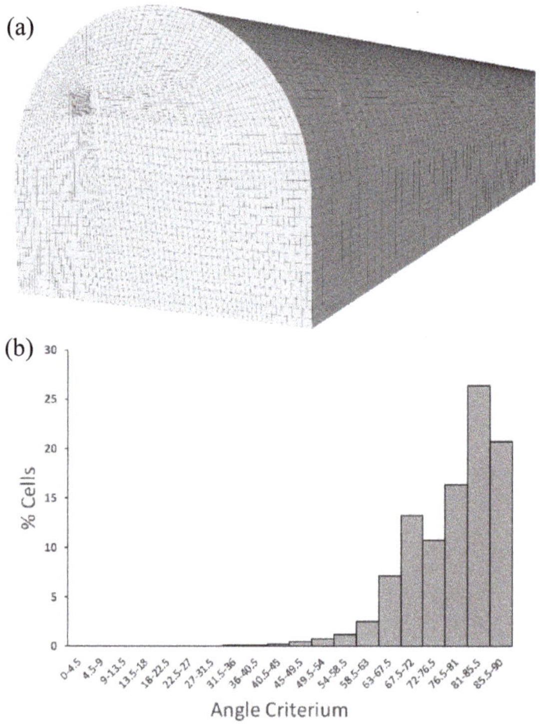

Figure 4. (a) Mesh and (b) mesh quality in terms of angle criterion.

In addition, a sensibility analysis of the mesh was also carried out. In Figure 5 the effective ventilation distance in front of the number of mesh cells in millions is shown. Three meshes were simulated (0.7 M/1 M/4 M cells) for scenario 2 and 20 m/s at the inlet boundary condition. It can be seen that the effective ventilation distance depends on the number of cells used, but the error committed with each of the meshes is negligible, as opposed to the value of the distance (about 1.5% for the mesh of 0.7 M cells). Nevertheless, in order to minimize the error while maintaining an optimum simulation time, the mesh of 1 million cells was selected for the subsequent experiments. The value for this mesh is near to the convergence value with a deviation of 0.3% (Figure 5), which is an assumable error.

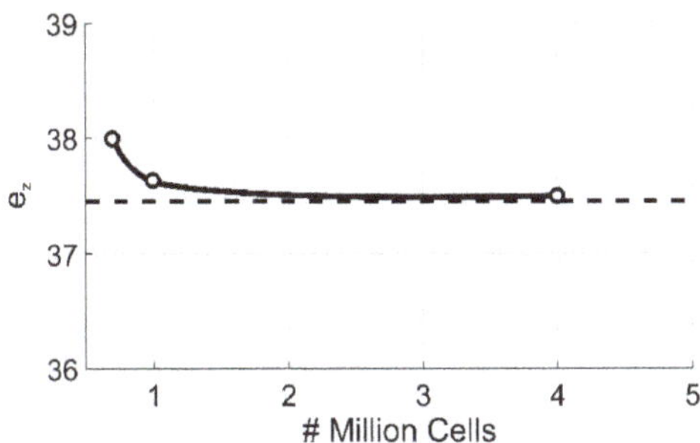

Figure 5. Mesh sensibility analysis.

2.3. CFDs Model

Since the Navier–Stokes equations cannot be solved analytically, the assessment of the flow pattern was performed by means of ANSYS Fluent software. This software solves the Navier–Stokes Equations (1)–(7) [23] using the finite volume model.

Continuity equation:

$$\frac{\partial \rho}{\partial t} + \nabla * \left(\rho * \vec{v_a} \right) = S_m.$$ (1)

Simplified to the problem:

$$\frac{\partial(\rho u)}{\partial x} + \frac{\partial(\rho v)}{\partial y} + \frac{\partial(\rho w)}{\partial z} = 0.$$ (2)

Momentum conservation equation:

$$\frac{\partial}{\partial t}\left(\rho \vec{v_a} \right) + \nabla * \left(\rho \vec{v_a} \vec{v_a} \right) = -\nabla P + \nabla * \left(\overline{\overline{\tau}} \right) + \rho \vec{g} + \vec{F}.$$ (3)

Simplified to the problem and decomposed in three axes:

$$-\frac{\partial P}{\partial x} + \frac{\partial(\tau xx)}{\partial x} + \frac{\partial(\tau yx)}{\partial y} + \frac{\partial(\tau zx)}{\partial z} = \nabla * \left(\rho \vec{u} \vec{u} \right),$$ (4)

$$-\frac{\partial P}{\partial y} + \frac{\partial(\tau xy)}{\partial x} + \frac{\partial(\tau yy)}{\partial y} + \frac{\partial(\tau zy)}{\partial z} = \nabla * \left(\rho \vec{v} \vec{u} \right),$$ (5)

$$-\frac{\partial P}{\partial z} + \frac{\partial(\tau xz)}{\partial x} + \frac{\partial(\tau yz)}{\partial y} + \frac{\partial(\tau zz)}{\partial z} = \nabla * \left(\rho \vec{w} \vec{u} \right).$$ (6)

The stress vector was calculated as:

$$\tau = \mu\left[\left(\nabla \vec{v_a} + \nabla \vec{v_a}^T \right) - \frac{2}{3} \nabla * \vec{v_a} I \right].$$ (7)

2.4. Turbulence Model

The equations above were solved using the SIMPLE scheme for the pressure–velocity coupling. In addition, for turbulence modeling, a k-ε model was selected [24]. The k-ε model is a two-equation model, which aims at resolving the turbulence using two parameters: the turbulence kinetic energy (k) and the dissipation rate (ε).

The equations solved by this model are listed as Equations (8) and (9) [23]:

$$\frac{\partial}{\partial t}(\rho k) + \frac{\partial}{\partial x_i}(\rho k u_i) = \frac{\partial}{\partial x_j}\left[\left(\mu + \frac{\mu_t}{\sigma_k}\right)\frac{\partial k}{\partial x_j}\right] + G_k + G_b - \rho\varepsilon - Y_M + S_k, \tag{8}$$

$$\frac{\partial}{\partial t}(\rho\varepsilon) + \frac{\partial}{\partial x_i}(\rho\varepsilon u_i) = \frac{\partial}{\partial x_j}\left[\left(\mu + \frac{\mu_t}{\sigma_\varepsilon}\right)\frac{\partial\varepsilon}{\partial x_j}\right] + C_{1\varepsilon}\frac{\varepsilon}{k}(G_k + C_{3\varepsilon}G_b) - C_{2\varepsilon}\rho\frac{\varepsilon^2}{k} + S_\varepsilon, \tag{9}$$

Where:

$$\mu_t = \rho C_\mu \frac{k^2}{\varepsilon}$$

$$C_{3\varepsilon} = \tanh\left|\frac{v}{u}\right|$$

2.5. Boundary Conditions

The boundary conditions are represented in Figure 6. Walls were defined following the default criteria of the software. The inlet velocity varied from 1 to 20 m/s according to the practical values offered by Vergne (2003). The velocity at the pressure outlet was set uniform at 0 Pa, so data were in manometric pressure.

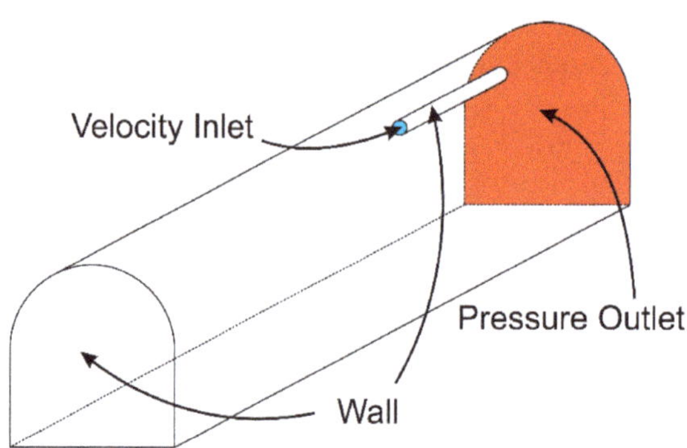

Figure 6. Boundary conditions of the simulations.

2.6. Software Configuration

In order to obtain reliable data, the following simulation parameters (Table 2) were selected.

Table 2. Simulation parameters.

Turbulence Model	**Realizable *k-ε* with Non-Equilibrium**
Pressure–Velocity Coupling	SIMPLE Scheme
Transient Formulation	Second-Order Implicit
Spatial Discretization	
Gradient	Green-Gauss Cell-Based
Pressure	Second Order
Momentum	Second Order
Turbulent Kinetic Energy	Second Order
Turbulent Dissipation Rate	Second Order

To minimize the results, errors all residuals of the simulation were set at 10^{-5}. This fact enables us to obtain more reliable data. Also, all simulations were initialized with 0 velocity in the three components, which is more important in the mixture model simulations.

2.7. Dead Zone Delimitation

The dead zone is the area in which the ventilating flow is not capable of sweeping the blasting gases. For practical reasons, several authors and legislations have defined the dead zone as that where the air velocity is at most 0.5 m/s [19] or, less restrictively, at most 0.15 m/s [18,25]. This definition, for a long period, implies that any air clearance is still possible until the air velocity is zero. We can see in the discussion section that this definition does not fit into reality.

For this reason, two alternative methodologies were tested in order to design the dead zone, which are: (a) air jet velocity analysis through the tunnel axis; (b) analysis of the area-weighted mean vertical speed in sections transverse to the tunnel axis.

For this first method, the control surfaces 1.25 times as large as the duct diameter and coaxial with it, were generated every half meter, covering the distance between the duct and the face. In the second approach, control surfaces were generated every half meter, covering the entire section of the tunnel.

2.8. Mixture Model

A mixture model is a simplified multiphase model regularly used to simulate homogeneously mixed flows if the coupling between the phases is strong. This model was selected to monitor the concentration of a tracer gas inside the dead-end tunnel [26]. In general, the most problematic gases after blasting are CO and NO_x ($NO + NO_2$), of which NO_2 is the most critical. This is because NO quickly oxidizes to NO_2, which is the most toxic and diffuses with greater difficulty.

All simulations were initialized at a concentration of 5% in the volume of this gas. Different flow rates of fresh air (0% NO_2) were simulated at the inlet to compare them with the aerodynamic simulation results. Ventilation times not greater than 20 min were considered because longer re-entry times after blasting significantly reduce tunneling productivity. All simulations were conducted for scenario 4 and with fresh air duct velocities less than 20 m/s.

3. Results and Discussion

3.1. Dead Zone Delimitation and Flow Field Patterns

Figure 7 was obtained by plotting contours of the air velocity magnitude in a longitudinal plane of the tunnel, which intersects the duct. It can be observed that the air jet tends to adhere to the tunnel roof owing to the Coandă effect [27,28]. This effect increases friction and energy losses of the air jet and reduces the effective distance. Moreover, as the jet flow expands, the air velocity decreases down to values close to zero. In light of this, the first option would be to consider the effective distance as that in which the air velocity is almost zero.

Figure 7. Velocity contours of the air jet exiting the ventilation duct along the tunnel axis. The data correspond to scenario 4 and 20 m/s air velocity at the velocity inlet.

Figure 8 depicts the average velocity of the air jet along the tunnel, measured in control surfaces 1.25 time as long as the duct diameter and coaxial with it. In this way, a continuum decrease in the absolute value of the air jet velocity can be observed. This curve reaches a point when velocity is almost zero and, once again, this could be defined as the beginning of the dead zone.

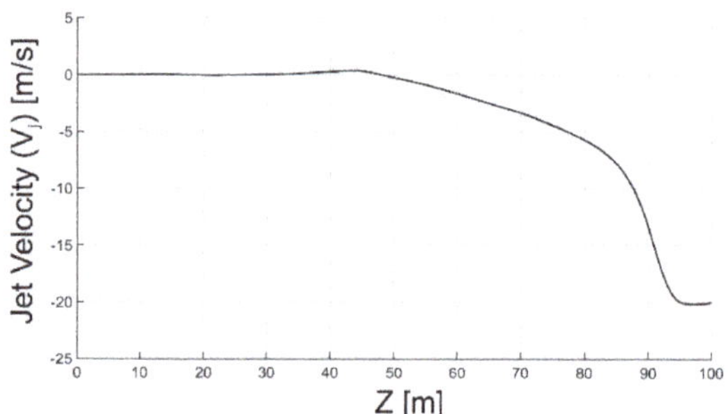

Figure 8. Jet velocity for scenario 4 and a duct velocity of 20 m/s.

Figure 9 shows that most part of the air jet changes its direction, returning through the gallery before entering the dead zone. The airflow in the dead zone is also dragged by the air jet nearby and generates counter vortices due to the shear stresses. As a result, the air continually re-circulates inside the dead zone without leaving it. This phenomenon invalidates the premise of defining the dead zone as just a low-velocity region.

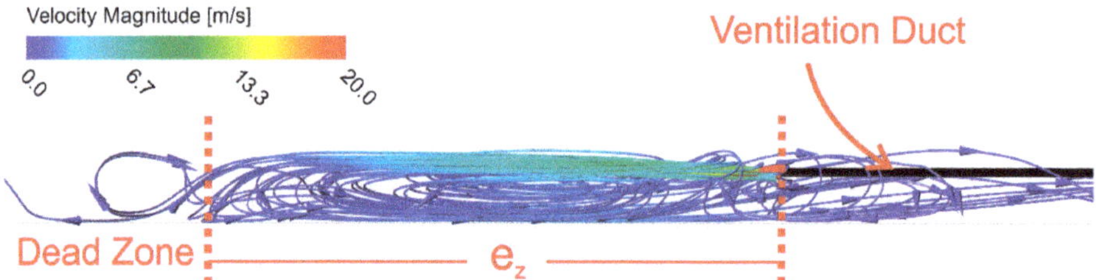

Figure 9. Counter vortices formation in the dead zone.

The analysis of the area-weighted average vertical air velocity (V_y) on control surfaces covering the entire transverse section of the tunnel is shown in Figure 10. The first phenomenon, observable at approximately 90 m from the face, was a zone of highly positive velocities. Upon analyzing the flow pattern, it can be perceived that judging from the simulation, the return flow is constricted in the lower part of the tunnel. However, at a certain moment the flow was free from the influence of the ventilation jet and filled the entire tunnel section. This phenomenon generates turbulences inside the tunnel and is responsible for this zone of mean positive vertical velocities in the simulation. Moreover, the negative mean vertical velocity observed in Figure 10 corresponds to the region where all the ventilation flow changes its direction towards the floor of the tunnel. This point is considered as the delimitation of the dead zone. In this way, we consider the limit of the dead zone where the negative vertical velocity reaches its maximum, although both approaches suggest very similar results.

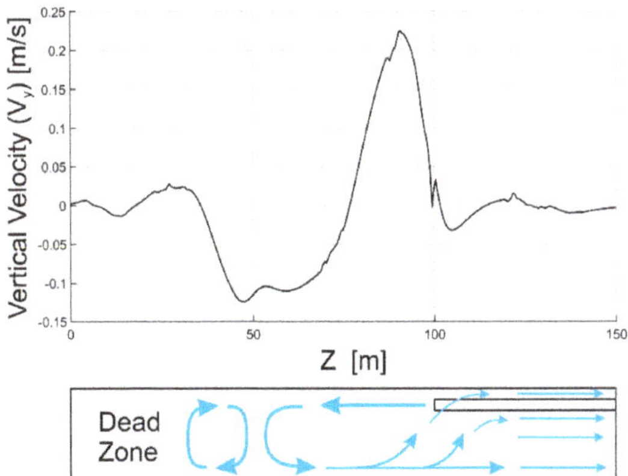

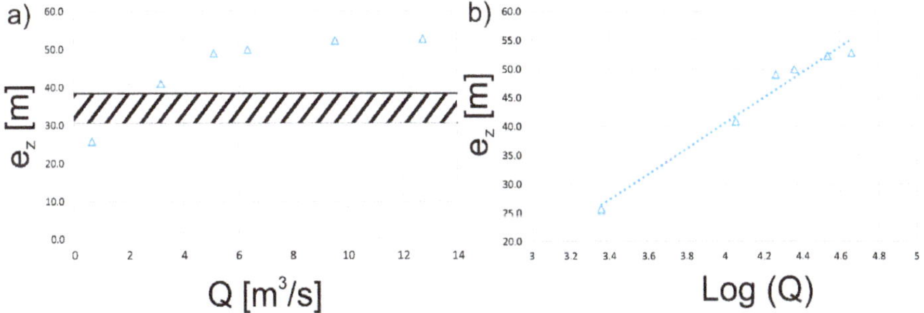

Figure 10. Mean vertical velocity along the tunnel axis for scenario 4 and inlet velocity 20 m/s.

Figure 11a shows the influence of the flow rate over the effective ventilation distance. According to this simulation, this distance grows rapidly for flow rates less than 1 m³/s, which are not common for mine ventilation purposes and reach a horizontal asymptote at a flow rate of about 5 m³/s. Moreover, this figure also shows a poor correlation between our expression and the legal requirements (shaded zone) for the low airflow rates and a good correlation for the large ones (horizontal asymptote). Moreover, it can be seen that the effective ventilation distance has a logarithmic relation to the flow rate. The linearization of this function allows us to assume that for common ventilation conditions, the effective distance is proportional to the logarithm of the flow rate (Figure 11b).

Figure 11. (a) Effective zone (e_z) versus flow rate (Q) for scenario 4. The shaded zone indicates the recommended distance ($4\sqrt{S}$ to $6\sqrt{S}$) between the duct end and the face; (b) the effective zone (e_z) versus the log flow rate (Q) for scenario 4.

3.2. Mathematical Model for the Effective Distance

Figure 12a shows the effective distance as a function of the flow rate for each scenario. Raw data were adimensionalized using the following coefficients, where the second coefficient was derived from the Reynolds number:

$$f_1 = \frac{e_z}{D_h},$$

$$f_2 = \frac{Q\rho}{D_h\mu},$$

where:

- e_z: effective distance (m);

- D_h: hydraulic diameter of the tunnel section (m);
- Q: ventilation flow rate (m/s);
- ρ: air density (kg/m^3);
- μ: air dynamic viscosity (Pa·s).

Figure 12b shows the transformed data according to the first dimensionless coefficient and the logarithm of the second. It was observed that if the curves were adimensionalized, they overlapped. That is to say, the solution is proportional if tunnel and duct sections remain constant. These results imply that this section alone should no longer be a parameter to study, and further work should be undertaken to study other parameters—For instance, the ratio between tunnel and duct sections. In general terms, it can be observed that the larger the tunnel section and the clean airflow rate, the longer the maximum effective distance. These results are in accordance with Feroze and Genc [29].

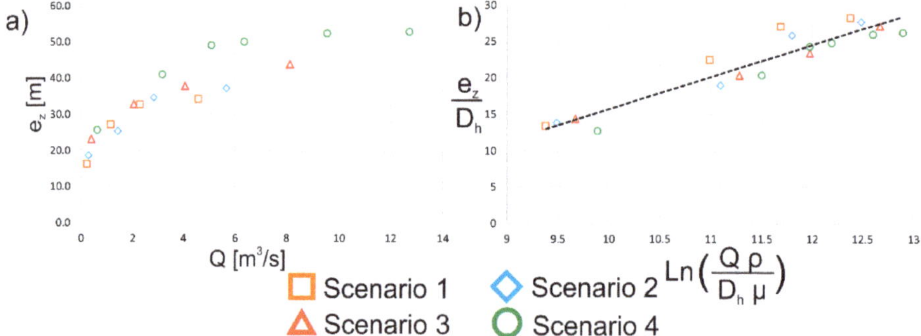

Figure 12. (**a**) Maximum effective distance (e_z) versus flow rate (Q); (**b**) dimensionless effective distance versus the logarithm of the dimensionless effective flow rate, for the different scenarios.

From the data in Figure 12b, a linear least squares (LLS) approximation was performed so that Equation (10) was obtained. This equation allows for the approximation of the effective distance as a function of the ventilation flow rate and the hydraulic diameter, which is defined as:

$$e_z = D_h\left(4.40\, Ln\frac{Q\rho}{\mu D_h} - 28.36\right) \tag{10}$$

$$D_h = \frac{4A}{P_w}$$

where:

- A: cross-sectional area of the tunnel (m^2);
- P_w: perimeter of the cross-section (m).

3.3. Mathematical Model Correction with the Mixture Model

The concentration of NO_2 along the tunnel axis for different ventilation times is shown in Figure 13. It can be seen that the majority of the gas dilutes in the first 12 min near the duct end (upper right part of the figure). This phenomenon corresponds to the aerodynamic sweep of NO_x by the clean air jet. Moreover, it can be observed that the clearance times are longer than 12 min. The NO_x concentration in the dead zone (left side of the figure) does not decrease significantly with the increasing ventilating time. This fact leads to the existence of a remaining concentration that can only be evacuated by diffusion. The minimum concentration remaining in the effective zone (bottom right of the figure) is the consequence of the existence of a dead zone, so the NO_x that exits this area by diffusion must cross the effective zone to leave the tunnel dead end, provoking a residual concentration of NO_x.

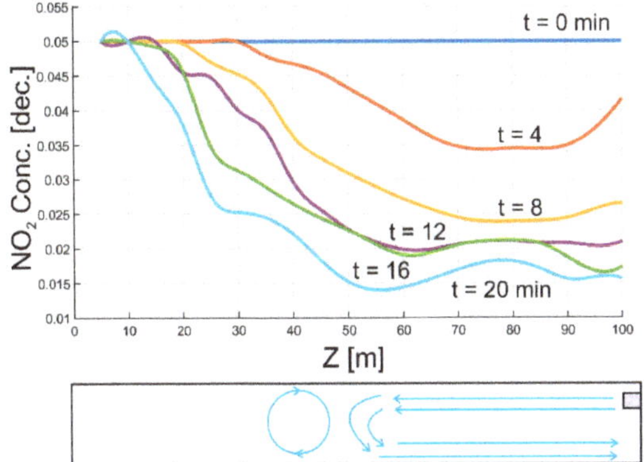

Figure 13. Concentrations of NO$_2$ inside the dead end for different ventilating times. The data correspond to scenario 4 and an air velocity of 20 m/s at the duct inlet.

The diffusion mechanism is not reliable or fast enough to be taken into account for tunnel ventilation purposes, so only the distance swept by the air and not the distance cleaned by diffusion has been considered for the determination of the final effective ventilation distance corrected by the mixture model. A comparison of the data derived from the aerodynamic and the mixture model simulations is given in Figure 14. According to the mixture model, the recommended distance between the duct-end and the face is shorter than that in the aerodynamic model because in this case the time needed to evacuate the gases is taken into consideration. Moreover, if the ventilation arrangement is one that does not permit the formation of a dead zone, as could be possible in practice, the ventilating time would be shorter due to the delay imposed by NO$_x$ diffusion inside the dead zone in the evacuation of the gases. Thus, the mixture model dilution times shown in Figure 13 are longer than the real ones as no dead zone is expected in a real tunnel ventilation situation.

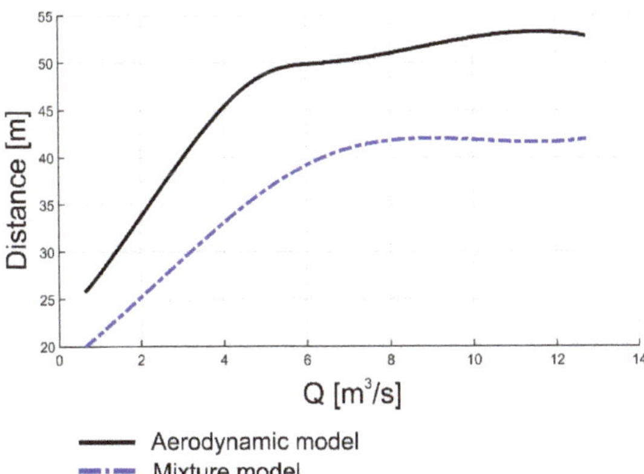

Figure 14. Maximum possible duct end-to-face distance for some ventilation of the dead zone, obtained with the aerodynamic and mixture models.

Finally, upon studying the relationship between the maximum distance recommended by the mixture and the aerodynamic models (Figure 14), it can be observed that this quotient is about 0.8. Thus, a coefficient, $C_1 = 0.8$, was added to Equation (10) and thus Equation (11) is obtained:

$$d_s = C_1 D_h \left(4.40 \, Ln \frac{Q\rho}{\mu D_h} - 28.36 \right) \tag{11}$$

It is important to keep in mind that Equation (11) does not search for the optimum distance for mine ventilation. On the contrary, it establishes a maximum duct end-to-face distance over which forced ventilation can evacuate blasting gases in a maximum of 20 min. Thus, these results do not disagree with the conventional recommendations that this distance should not to exceed (4 $\sqrt{S}$–6 $\sqrt{S}$), but longer clearance times should probably be expected with this expression (e.g., [19]).

Equation (11) is plotted for practical reasons in Figure 15. This graph allows for the approximation of the maximum duct end-to-face distance for some ventilation in terms of the tunnel section, clean airflow rate and ventilation time, taking into account the mixture model. Firstly, it can be observed that the larger the tunnel section, the longer the effective distance. This fact agrees with legal requirements. Furthermore, our expression takes into consideration the clean airflow rate sweeping the face, which is not considered by legal specifications. In this way, our work indicates that the larger the airflow rate, the longer the effective distance. Better correlations between legal requirements and our expression were obtained for lower airflow rates, in the case of the smaller sections. On the contrary, less disparity was achieved in the case of higher airflow rates for the larger sections.

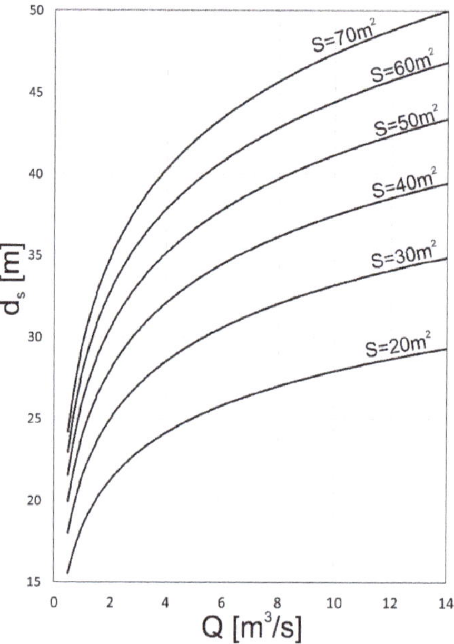

Figure 15. Maximum possible duct end-to-face distance for some ventilation of the dead zone, in terms of airflow rate and tunnel section.

4. Conclusions

The main aims of drivage ventilation are the evacuation of the blasting and machinery gases as well as the supply of the required air for the workers' breathing. This type of ventilation is usually performed by means of a forcing system. In this study, a full-scale three-dimensional CFDs model of a forcing system was created to assess the ventilation performance of a dead-end gallery. The simulation was performed for different parameters, namely, tunnel section, flow rate and position with regard to the tunnel axis.

The existence of a dead zone between the duct end and the face influences the ventilation system setup. This zone has been traditionally defined as a region of reduced velocity of the air current that hinders the mixture of gases with fresh air. Our study suggests that common approaches defining the dead zone as the region in which air velocity tends to zero are not accurate. This fact is consequence of the formation counter vortexes that recirculate gases inside this zone obstructing their evacuation. Thus, although very little air renovation takes place in this zone, the gases inside it are actually in motion.

As a result of our work, we obtained an expression to approximate the maximum effective ventilating distance for some ventilation as a function of the tunnel section and the ventilation flow rate, considering a CFDs analysis with a mixture model. The equation obtained should be valid for any tunnel of similar geometry. Our calculations indicate that the maximum effective distance is larger than that suggested by traditional legal requirements, thus confirming their level of conservativism. The results obtained may assist practicing engineers in improving ventilation productivity during tunnel construction works. Future research should also consider dust when establishing an effective duct end-to-face distance for dead-end tunnel ventilation.

Author Contributions: M.G.-D. improved the methodology, performed the simulations and co-wrote the original draft. C.S. conceptualized the problem, analyzed the results and co-wrote the original draft. C.M.-G. reviewed and edited the draft manuscript, improved visualization and validated the data. B.P. performed the formal analysis, provided the resources and acquired founding. All authors discussed the results and contributed to the final manuscript.

Funding: This research received no external funding.

Acknowledgments: M. G.-D. was supported by the Spanish "Ministerio de Educación, Cultura y Deporte" within the "FPU" Program (grant number FPU15/04375). C.M.G. was supported by the Spanish "Ministerio de Educación Cultura y Deporte" within the "Doctorados Industriales" Program (grant number DI-17-09596).

Conflicts of Interest: No potential conflict of interest was reported by the authors.

Nomenclature

$C_{1\varepsilon}$ Constant (-) (value 1.44)
$C_{2\varepsilon}$ Constant (-) (value 1.92)
$C_{3\varepsilon}$ Constant (-)
C_μ Constant (-) (value 0.09)
F Model-dependent source terms (Pa m^{-1})
G_b Generation of k due to buoyancy (m^2 s^{-2})
G_k Generation of k due to mean velocity gradients (m^2 s^{-2})
I Identity matrix (-)
k Turbulent kinetic energy (m^2 s^{-2})
P Pressure (Pa)
S Tunnel section (m^2)
S_k User-defined source terms for k (m^2 s^{-2})
S_m Sources of mass (kg m^{-3} s^{-1})
S_ε User-defined source terms for ε (m^2 s^{-2})
t Time (s)
u velocity in x axis (m s^{-1})
v velocity in y axis (m s^{-1})
v_a Absolute velocity (m s^{-1})
w velocity in z axis (m s^{-1})
Y_M Contribution of the fluctuating dilatation in compressible turbulence (m^2 s^{-2})
Greek symbols
$\overline{\overline{\tau}}$ Stress tensor (Pa)
ε Dissipation rate (m^2 s^{-3})
μ Dynamic viscosity (Pa s)
μ_t Tubulent viscosity (Pa s)
ρ Density (kg m^{-3})
σ_k Prandtl number for k (-) (value 1.0)
σ_ε Prandtl number for ε (-) (value 1.3)

References

1. Sarmah, M.; Khare, P.; Baruah, B.P. Gaseous emissions during the coal mining activity and neutralizing capacity of ammonium. *Water Air Soil Pollut.* **2012**, *223*, 4795–4800. [CrossRef]
2. Osunmakinde, I.O. Towards safety from toxic gases in underground mines using wireless sensor networks and ambient intelligence. *Int. J. Distrib. Sens. Netw.* **2013**, *2013*, 159273. [CrossRef]
3. Vutukuri, V.S.; Lama, R.D. *Environmental Engineering in Mines*; Cambridge University Press: Cambridge, UK, 1986.
4. Borísov, S.; Klókov, M.; Gornovói, B. *Labores Mineras*; Mir: Moscú, Russia, 1976.
5. Novitzky, A. *Ventilación de Minas*; L'autor: Buenos Aires, Argentina, 1962.
6. Kissell, F. *Handbook for Methane Control in Mining*; DHHS (NIOSH): Pittsburg, PA, USA, 2006; Volume 131.
7. Lu, Y.; Akhtar, S.; Sasmito, A.P.; Kurnia, J.C. International Journal of Mining Science and Technology Prediction of air flow, methane, and coal dust dispersion in a room and pillar mining face. *Int. J. Min. Sci. Technol.* **2017**, *27*, 657–662.
8. Ren, T.; Wang, Z.; Cooper, G. CFD modelling of ventilation and dust flow behaviour above an underground bin and the design of an innovative dust mitigation system. *Tunn. Undergr. Space Technol.* **2014**, *41*, 241–254. [CrossRef]
9. Cheng, J.; Li, S.; Zhang, F.; Zhao, C.; Yang, S.; Ghosh, A. CFD modelling of ventilation optimization for improving mine safety in longwall working faces. *J. Loss Prev. Process Ind.* **2016**, *40*, 285–297. [CrossRef]
10. Lolon, S.A.; Brune, J.F.; Bogin, G.E.; Grubb, J.W.; Saki, S.A.; Juganda, A. Computational fluid dynamics simulation on the longwall gob breathing. *Int. J. Min. Sci. Technol.* **2017**, *27*, 185–189. [CrossRef]
11. Sasmito, A.P.; Kurnia, J.C.; Birgersson, E.; Mujumdar, A.S. Computational evaluation of thermal management strategies in an underground mine. *Appl. Therm. Eng.* **2014**, *90*, 1144–1150. [CrossRef]
12. Yuan, L.; Smith, A.C. *Computational Fluid Dynamics Modeling of Spontaneous Heating in Longwall Gob Areas*; CDC Stacks; Publick Health Publications: Atlanta, GA, USA, 2002.
13. Shi, G.Q.; Liu, M.; Wang, Y.M.; Wang, W.Z.; Wang, D.M. Computational Fluid Dynamics Simulation of Oxygen Seepage in Coal Mine Goaf with Gas Drainage. *Math. Probl. Eng.* **2015**, *2015*, 723764. [CrossRef]
14. Diego, I.; Torno, S.; Toraño, J.; Menéndez, M.; Gent, M. A practical use of CFD for ventilation of underground works. *Tunn. Undergr. Space Technol. Inc. Trenchless Technol. Res.* **2011**, *26*, 189–200. [CrossRef]
15. Toraño, J.; Torno, S.; Menéndez, M.; Gent, M. Auxiliary ventilation in mining roadways driven with roadheaders: Validated CFD modelling of dust behaviour. *Tunn. Undergr. Space Technol.* **2011**, *26*, 201–210. [CrossRef]
16. Sasmito, A.P.; Birgersson, E.; Ly, H.C.; Mujumdar, A.S. Some approaches to improve ventilation system in underground coal mines environment—A computational fluid dynamic study. *Tunn. Undergr. Space Technol.* **2013**, *34*, 82–95. [CrossRef]
17. Toraño, J.; Torno, S.; Menendez, M.; Gent, M.; Velasco, J. Models of methane behaviour in auxiliary ventilation of underground coal mining. *Int. J. Coal Geol.* **2009**, *80*, 35–43. [CrossRef]
18. Li, M.; Aminossadati, S.M.; Wu, C. Numerical simulation of air ventilation in super-large underground developments. *Tunn. Undergr. Space Technol.* **2016**, *52*, 38–43. [CrossRef]
19. Fang, Y.; Yao, Z.; Lei, S. Air flow and gas dispersion in the forced ventilation of a road tunnel during construction. *Undergr. Space* **2018**, *4*, 168–179. [CrossRef]
20. Szlazak, N.; Szlazak, J.; Tor, A.; Obracaj, A.; Borowski, M. Ventilation systems in dead end headings with coal dust and methane hazard. In Proceedings of the 30th International Conferences of safety in Mines Research Institutes, Johannesburg, South Africa, 5–9 October 2003.
21. Reed, W.; Taylor, C. Factors affecting the development of mine face ventilation systems in the 20th century. In Proceedings of the Society for Mining, Metallurgy and Exploration (SME): Annual Meeting and Exhibit, Denver, CO, USA, 25–28 February 2007.
22. De La Vergne, J.; Mcintosh, S. *Hard Rock Miners Handbook*; Stantec: Edmonton, AB, Canada, 2000; Volume 4, ISBN 0-9687006-0-8.
23. ANSYS. *ANSYS FLUENT 16.2 User's Guide*; ANSYS: Canonsburg, PA, USA, 2016.
24. López González, M.; Galdo Vega, M.; Fernández Oro, J.M.; Blanco Marigorta, E. Numerical modeling of the piston effect in longitudinal ventilation systems for subway tunnels. *Tunn. Undergr. Space Technol.* **2014**, *40*, 22–37. [CrossRef]

25. Ministry of Water Resources of the People's Republic of China. *Construction Specifications on Underground Excavation Engineering of Hydraulic Structures*; China Water Power Press: Beijing, China, 2007.

26. Kurnia, J.C.; Sasmito, A.P.; Mujumdar, A.S. Simulation of a novel intermittent ventilation system for underground mines. *Tunn. Undergr. Space Technol.* **2014**, *42*, 206–215. [CrossRef]

27. Betta, V.; Cascetta, F.; Musto, M.; Rotondo, G. Numerical study of the optimization of the pitch angle of an alternative jet fan in a longitudinal tunnel ventilation system. *Tunn. Undergr. Space Technol.* **2009**, *24*, 164–172. [CrossRef]

28. Ji, J.; Fan, C.G.; Gao, Z.H.; Sun, J.H. Effects of Vertical Shaft Geometry on Natural Ventilation in Urban Road Tunnel Fires. *J. Civ. Eng. Manag.* **2014**, *20*, 466–476. [CrossRef]

29. Feroze, T.; Genc, B. Analysis of the effect of ducted fan system variables on ventilation in an empty heading using CFD. *J. S. Afr. Inst. Min. Metall.* **2017**, *117*, 157–167. [CrossRef]

© 2019 by the authors. Licensee MDPI, Basel, Switzerland. This article is an open access article distributed under the terms and conditions of the Creative Commons Attribution (CC BY) license (http://creativecommons.org/licenses/by/4.0/).

Article

Life Cycle Environmental Costs of Buildings

Yuanfeng Wang [1,*], Bo Pang [1], Xiangjie Zhang [1], Jingjing Wang [2], Yinshan Liu [1], Chengcheng Shi [1] and Shuowen Zhou [3]

[1] School of Civil Engineering, Beijing Jiaotong University, Beijing 100044, China; 11115277@bjtu.edu.cn (B.P.); q276008525@163.com (X.Z.); 16121069@bjtu.edu.cn (Y.L.); 17115326@bjtu.edu.cn (C.S.)
[2] College of Architecture and Civil Engineering, Beijing University of Technology, Beijing 100124, China; 18811442841@163.com
[3] China Academy of Building Research, Beijing 100013, China; zhoushuowen@126.com
* Correspondence: cyfwang@bjtu.edu.cn; Tel./Fax: +86-10-51685552

Received: 21 February 2020; Accepted: 12 March 2020; Published: 14 March 2020

Abstract: Energy consumption and pollutant emissions from buildings have caused serious impacts on the environment. Currently, research on building environmental costs is quite insufficient. Based on life cycle inventory of building materials, fossil fuel and electricity power, a calculating model for environmental costs during different stages is presented. A single-objective optimization model is generated by converting environmental impact into environmental cost, with the same unit with direct cost. Two residential buildings, one located in Beijing and another in Xiamen, China, are taken as the case studies and analyzed to test the proposed model. Moreover, data uncertainty and sensitivity analysis of key parameters, including the discount rate and the unit virtual abatement costs of pollutants, are also conducted. The analysis results show that the environmental cost accounts for about 16% of direct cost. The environmental degradation cost accounts for about 70% of the total environmental cost. According to the probabilistic uncertainty analysis results, the coefficient of variation of material production stage is the largest. The sensitivity analysis results indicate that the unit virtual abatement cost of CO_2 has the largest influence on the final environmental cost.

Keywords: building; environmental costs; green GDP, China; uncertainty analysis; sensitivity analysis

---

1. Introduction

With the rapid development of economy, China overtook the US as the world's biggest energy consumer and greenhouse gas (GHG) emitter. About 1.6–2.0 billion m^2 of buildings are constructed every year in China [1], accounting for about 40% of the world's total new buildings [2]. A large amount of GHG will be emitted during the life cycle of buildings, especially in construction and operation stages. In order to achieve the sustainable development of construction, there is a great need to clearly know both the costs and the environment costs of buildings.

At present, there is no common understanding of the concept of environmental cost in the academic circle, and there are still some differences among different research fields. According to United States Environmental Protection Agency (USEPA) [3], how environmental costs are defined depends on how the information is used. Whether a cost can be defined as environmental cost is not absolute but needs to be considered according to specific research purpose. The definition of environmental costs is more representative in the System of Integrated Environmental and Economic Accounting (SEEA) published by the United Nations Statistics Division (UNSD) in 1993 [4]. According to the definition, environmental costs consist of two levels: (1) the use and loss value of natural resources in output and final consumption; (2) the impact value of pollution generated by output and consumption activities on environment. In addition, the United States Council on Environmental Quality divides environmental costs into four parts: environmental loss costs, environmental protection costs, environmental affairs costs and environmental pollution elimination costs.

In China, the Research Group on Integrated Environmental and Economic Accounting (Green GDP) proposed in its technical guidance that environmental costs are composed of pollution control costs and environmental degradation costs, among which pollution control costs can be divided into actual pollution control costs and virtual pollution control costs. Based on the definition, the Research Group has conducted a study on China's green national economic accounts and published a number of studies on China's environmental economic accounts [5–7].

The relationship between environmental performance and economic performance is critical for environmental cost analysis. Several methodologies have been proposed to reveal the relationship, such as life cycle cost analysis (LCCA), whole life cost, eco-cost and eco-efficiency. Usually, the LCCA term implies that environmental costs are not included, as is the case in the similar whole life cost. Eco-efficiency has been proposed as one of the main tools to facilitate the transformation from unsustainable developments to sustainable developments [8]. It is based on the concept of increasing productivity and reducing economic and environmental performance at the same time [9,10]. Eco-efficiency refers to the ratio between the added value of a product (e.g., GDP) and the environmental impacts of the product or service (e.g., SO_2 emissions) [9,11]. It has significant implications for environmental management accounting (EMA) system as well as environmental accounting [10,12].

Additionally, the environmental costs or eco-cost indicators are used to assess the environmental costs. Eco-costs are a measure to prevent the burden of products by expressing the amount of environmental burden. Vogtländer et al. [13] used "eco-costs 2007", an indicator for assessing ecosystem deterioration and human health problems, to compare the environmental impact of bamboo materials with commonly used materials such as timber. Baeza-Brotons et al. [14] applied eco-costs to evaluate the environmental impacts of cement with and without addition of sewage sludge ash. Kravanja and Čuček [11] presented a novel indicator called eco-profit, which was defined as the sum of eco-benefit (positive impact of environmental unburdens) and eco-cost (negative impact of environmental burdens).

For the application of environmental cost in civil engineering, only handful of studies can be found. Kendall et al. [15] proposed an integrated life cycle assessment (LCA) and LCCA model to assess and compare traditional concrete bridges with cement-based composite bridges. The LCCA they calculated includes construction, consumer and environmental costs, reflecting the loss caused by air pollution. Chen [16] established a life cycle environmental impact cost analysis index system of bridges based on LCCA, calculating life cycle environmental costs of bridges at different stages. The results show that, among all stages, the environmental cost of the material production stage is higher than that of any other stages. A method translating the environmental impact into monetary units was composed by Carreras et al. [17]. The approach used eco-cost indicators to quantify the cost to prevent a given amount of environmental burden. However, the eco-costs only considered the material consumption and energy consumption. Chou and Yeh [18] developed a CO_2 emissions evaluation system and an environmental cost calculation method to compare the difference of environmental performance between fully prefabricated and cast-in-situ construction. In their study, CO_2 emissions were simply converted into environmental costs by referencing the profit-seeking enterprise income tax in Taiwan, and the progressive tax rate was used to transform the simulated total CO_2 emissions into environmental cost.

Through literature review, studies of building environmental costs, especially the life cycle environmental costs, are still quite insufficient. At present, several existing issues could complicate these efforts in research on environmental cost in civil engineering. For instance, environmental costs are always underestimated. Additionally, lack of adequate measuring and managing systems of environmental costs is another obstacle [19]. To overcome this gap, a calculating model for environmental costs of a building throughout life cycle is presented in this paper to obtain total energy consumption and pollutant emission costs of buildings.

The aim of this paper is to establish a single-objective optimization model by converting environmental impact into environmental cost, with the same unit of direct cost. The following investigations are conducted: (1) Firstly, this study builds an LCA model with all processes; (2) A virtual abatement cost of pollutants and environmental degradation cost according to macroscopic data of environmental economic accounting in China is calculated; (3) The green construction measures fee is incorporated into the environmental cost for the characteristics of building construction; (4) In order to analyze the differences in northern and southern parts of China, two residential buildings, one located in Beijing and the other in Xiamen, China, are taken as case studies; (5) Uncertainty analysis is carried out, including model and data uncertainties to evaluate how these sources of uncertainty may affect the environmental cost results; (6) Finally, sensitivity analysis of the environmental costs is conducted to identify major input variables, including the discount rate and the unit virtual abatement costs of pollutants.

2. Methods

According to the Guideline for Chinese Environmental and Economic Accounting [5] and the characteristics of construction engineering, the environmental costs of buildings are divided into three parts: (1) green construction measures cost, which refers to the practical costs of protecting the environment during construction stage; (2) virtual abatement costs, which are used to control the emissions of pollutants in the life cycle of buildings, including water pollutants, air pollutants and solid waste pollutants, and where C_{va1}, C_{va2} and C_{va3} are the virtual abatement costs of air pollution, water pollution and solid waste pollution, respectively; (3) environmental degradation cost, which is the environmental loss cost caused by the emission and pollution of buildings, where C_{ed1}, C_{ed2} and C_{ed3} are the environmental degradation costs of air pollution, water pollution and solid waste pollution, respectively.

The flowchart of the model is demonstrated in Figure 1. Based on the collected project inventory, three types of environmental pollution including air, water and solid waste pollution will be quantified. Based on the quantified results and the methods proposed in this paper, the virtual abatement costs and environmental degradation costs of the three types of environmental pollution can be obtained. Finally, the total environmental costs of a building will be obtained by adding green construction measures costs of subengineering fees including construction, decoration and erection works.

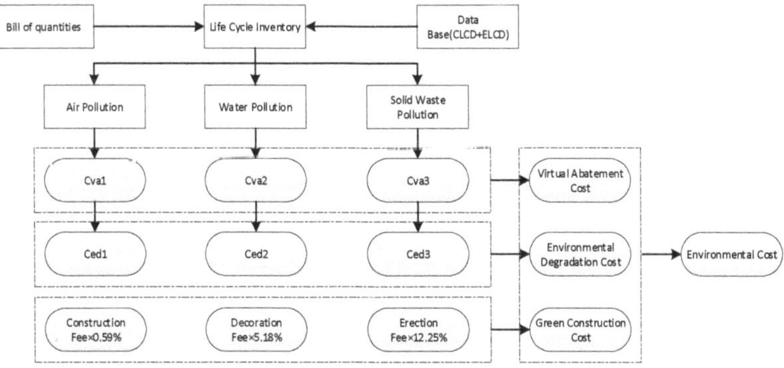

Figure 1. The flowchart of the model.

2.1. Green Construction Measures Cost

The Chinese government has proposed to levy green construction measures costs to improve the energy efficiency of construction. Green construction measures cost (C_{gc}) refers to environmental protection fees, which are used to reduce the negative impact of construction and consumption of resources under the condition of ensuring engineering quality and safety. The ratio of green construction measure costs to subengineering fees of an actual engineering project is shown in Table 1.

Table 1. Ratio of green construction measures cost to subengineering fees.

-	Construction Work (%)	Decoration Work (%)	Erection Work (%)
Resource conservation	0.20	2.05	2.25
Environmental protection	0.10	0.86	1.75
Health and safety	0.29	2.27	8.25
Total	0.59	5.18	12.25

2.2. Virtual Abatement Costs

The virtual abatement cost represents the cost of curbing untreated environmental pollutants. Three pollutants categories are included: water pollutants (including COD and ammonia), air pollutants (including SO_2, dust, fine particulate matter and NO_x) and solid waste pollutants (including household waste in operation stage and building material waste in demolition stage). The virtual abatement cost is calculated based on the quantity of pollutant emissions, i.e., the results of life cycle inventory, and virtual abatement costs of per unit pollutant, which is in accordance with the Guideline for Chinese Environmental and Economic Accounting.

2.2.1. Life Cycle Inventory

The framework selected in this study is in the light of the standards of ISO [20] and the Society of Environmental Toxicology and Chemistry (SETAC) [21]. The functional unit is considered as floor area (m^2). The cut-off principle of this study is in reference to previous research [22]: sorting all the building materials according to their mass, with the cumulative quality accounting for more than 80% of the building materials being taken into consideration.

As the two case studies are located in China, a local LCI database, Chinese Life Cycle Database (CLCD), is preferred. Although the life cycle inventory (LCI) has achieved remarkable process since last decade, the local LCI database is not able to cover all the material. Therefore, the Europe Life Cycle Database (ELCD) [23] is used to complete the case studies (see Table 2).

Table 2. Data sources of the study.

Life Cycle Stage	Subprocess	Data Origin
Material production	Concrete C20	ELCD
-	Concrete C25	ELCD
-	Concrete C30	CLCD
-	Concrete C35	ELCD
-	Concrete C50	CLCD
-	Rebar	CLCD
-	Rolled section steel	CLCD
-	Steel tube	CLCD
-	Cement mortar	ELCD
-	Wood door	ELCD
-	Aluminum door	ELCD
-	Window frame, Aluminum	ELCD
-	Alkyd paint	ELCD
-	Glass curtain wall	CLCD
Construction	Diesel	CLCD
-	Gasoline	CLCD
-	Electricity	CLCD
Operation	Water	CLCD
-	Electricity	CLCD
-	Natural gas	CLCD

- Material production stage

Pollutant emissions produced in this stage can be calculated based on the bill of material quantities and the life cycle inventory. A proper material loss rate has been considered in the bill of quantities, which references the Quota of Beijing Construction Project [24].

- Construction stage

The two main sources of pollutant emissions produced in this stage are construction machines and material transportation. Gasoline, diesel and electricity consumed by construction machines are calculated based on National Unified Construction Machinery Quota [25]. In the light of 2013 Statistical Yearbook of China, the average transportation distance is 181 km [2]. It is assumed that building materials are transported by trucks. The average fuel consumption level is about 101.78 L/(kt·km) [13]. The diesel consumption can be calculated as follows:

$$Q = \sum_i m_i \times L_i \times q_{m_i} \tag{1}$$

where Q is the diesel consumption; m_i is the mass of i-th material; L_i is the transportation distance of i-th material, assumed to be 181 km; q_{mi} is the average fuel consumption for transporting per unit material, assumed to be 101.78 L/(kt·km).

- Operation stage

Energy consumption during this stage implicates the energy and resources, including electricity, natural gas and water consumption. Since the two case buildings just completed construction, there are no actual maintenance monitoring data. Consequently, the water consumption, electricity consumption and domestic waste production for each person can only be estimated based on the local statistical yearbook [26,27], assuming that each family consists of three people. The number of apartments in the two case study buildings is 78 for Xiamen and 100 for Beijing.

Additionally, the pollutant emissions also include household waste, which can be estimated based on household waste of similar commercial buildings per unit time. For residential buildings, the energy consumption and household waste amount are influenced by per capita consumption and living habits, which can be estimated in the light of the statistical yearbook. For regions in northern China, the environmental costs caused by the consumption of coal for heating cannot be ignored.

- Demolition stage

The data about energy consumption of China's construction in the demolition stage are very scarce. The percentages of landfill, incineration and recycling in this paper are based on the data provided by Fabre [28], Zeng [29] and Lei et al. [30], who collected the current inventory data of construction waste recycle and landfill, mainly considering the resource consumption during recycle and landfill. The inventory data of construction waste is shown in Table 3.

Table 3. The inventory data of construction waste (unit: g/t).

-	Oil	Coal	Iron	Limestone
Recycle	3931	394	105	119
Reuse	4588	460	180	204
Landfill	2342	234	15	17

2.2.2. Virtual Abatement Costs of Pollutants

The virtual abatement costs (C_{va}) generated by the air and water pollution generated during the building life cycle can be quantified based on the bill of quantities and life cycle inventory. The formula is as follows:

$$C_{va1} + C_{va2} = Q_1 \times c_{va1} + Q_2 \times c_{va2} \tag{2}$$

where Q_1 is the amount of air pollutants, based on LCI; Q_2 is the amount of water pollutants, based on LCI; c_{va1} is the unit virtual abatement costs of air pollution (see Table 4); c_{va2} is the unit virtual abatement costs of water pollution (see Table 5).

Table 4. Unit virtual abatement costs of air pollution (unit: CNY/t).

-	CO_2	CO	CH_4	SO_2	NO_x	Dust	N_2O	VOC
Unit virtual costs	140.07	13.34	2561.26	650	3030	140	47,437.04	13,073.2

Table 5. Unit virtual abatement costs of water pollution (CNY/kg).

Industry	Steel	Manufacturing Industry	Electricity	Fuel Gas
COD	1.4	8.4	15.3	2.8
NH_4^+	2.291	0.039	0.516	2.001

2.2.3. Virtual Abatement Costs of Solid Waste

The solid waste produced in the building life cycle is composed of building solid waste and household waste.

- Building solid waste

The recycle rate of building material in China is considerable low. Most of building solid waste is simply treated by depositing or burying in the suburb, which will cause severe environmental pollution during transportation and deposition [31]. The abatement costs of building solid waste can be calculated as follows:

$$C_{va31} = Q_{31} \times c_{va31} \tag{3}$$

where C_{va31} is the virtual abatement costs of building solid waste; Q_{31} is the total amount of building solid waste; c_{va31} is the virtual abatement cost per unit building solid waste. According to the results of pollution loss survey data and System of Integrated Environmental and Economic Accounting (SEEA) of pilot provinces, the general industrial solid waste per unit virtual management cost is 22 CNY/t [5].

- Household waste

With the development of China's urbanization, most of household waste is disposed after harmless treatment, instead of directly drained off into the natural environment. The definition of harmless disposal is when advanced technology and scientific technology are used in the treatment of municipal solid waste to reduce the environmental impact of solid waste [32]. There are mainly three kinds of garbage harmless treatments: landfill, compost and incineration.

With the promotion of household waste treatment technology, some cities have achieved 100% harmless treatment. In this study, it is assumed that no harm will be caused by household waste after harmless treatment, and the environmental degradation costs can be ignored. The virtual abatement costs of household waste can be calculated as:

$$C_{va32} = Q_{32} \times c_{va32} + \sum_{k=1}^{n} Q_k \times c_{vak} \tag{4}$$

where C_{va32} is the virtual abatement costs of household waste; Q_{32} is the total amount of household waste; c_{va32} is the transportation costs of household waste; Q_k is the amount of household waste treated by different technologies; c_{vak} is the unit virtual abatement costs of each treatment (shown in Table 6).

Table 6. Unit virtual abatement cost of household waste.

Transportation	Sanitary Landfill	Innocent Treatment	Simple Landfill
25	35	60	8

By summing up C_{va1}, C_{va2} and C_{va3}, the total virtual abatement costs of building can be calculated as follows:

$$C_{va} = C_{va1} + C_{va2} + C_{va31} + C_{va32} \tag{5}$$

2.3. Environmental Degradation Costs

Environmental degradation cost (C_{ed}) indicates the economic value loss caused by the degradation of environmental functions. The environmental degradation cost is calculated by the pollution loss cost method. The pollution loss cost method requires a specific technical approach to conduct a special survey of pollution losses to determine the monetary value of the impact of pollution emissions on local environmental quality. After quantifying these influences, the environmental degradation costs caused by pollution can be determined.

The Chinese government published the Chinese Environmental and Economic Accounting Report 2004 [6]. As some local governments firmly opposed publishing the report, after 2008, there are no updated data that can be used to estimate environmental degradation costs.

In order to estimate the environmental degradation costs, a formula was established in the light of the ratio of environmental degradation costs to virtual abatement costs, shown as:

$$C_{ed} = \sum C_{vai} \times r_i \tag{6}$$

where C_{ed} is the total environmental degradation costs; C_{vai} is the virtual abatement costs of air pollution if $i = 1$, or water pollution if $i = 2$, or solid waste pollution if $i = 3$; r_i is the average ratio of environmental degradation costs to virtual abatement costs, according to the Chinese Environmental and Economic Accounting Report 2004 (see Table 7), $r_1 = 2.25$, $r_2 = 1.32$, $r_3 = 0.31$.

Based the discussion above, the total life cycle environmental costs can be calculated as:

$$C_e = C_{gc} + C_{va} + C_{ed} \tag{7}$$

where C_e is the total life cycle environmental costs; C_{gc} is the green construction measures costs; C_{va} is the virtual abatement costs; C_{ed} is the environmental degradation costs.

Since the time value of money concerns the effect of time and interest rate on monetary amounts, this effect must be given primary consideration in environmental cost [33]. Present value, also known as present discounted value, is the value of an expected income stream determined at the valuation date. The present value is always less than or equal to the future value due to the potential of interest-earning, which referred to as the time value of money. The most commonly applied model of present valuation uses compound interest.

The present value of the total environmental costs of a building can be expressed as:

$$C_{epv} = C_{ep} + \frac{C_{ec}}{t_1} \times (P|A, r, t_1) + C_{eo} \times (P|A, r, t_2) \times (P|A, r, t_1) + \frac{C_{edem}}{(1 + r)^{(t_1 + t_2)}} \tag{8}$$

where C_{epv} is the present value of the total environmental costs; C_{ep} is the environmental cost of the material production stage; C_{ec} is the environmental cost of the construction stage; C_{eo} is the environmental cost of the operation stage; C_{edem} is the environmental cost of the demolition stage; t_1 is the number of annual interest periods during construction stage, assumed to be 2 years; t_2 is the number of annual interest periods during operation stage, assumed to be 50 years; r is the discount rate, assumed to be 7%; A is the equal annual payment; $(P|A, r, t_i)$ is the equal-payment-series present-worth factor at time t_i, calculated as $(P|A, r, t_i) = A[(1+r)^{t_i} - 1]/r(1+r)^{t_i}]$.

Table 7. The accounting result of environmental costs in China from 2004–2008.

Air Pollution (Unit: Hundred Million Yuan)				
Year	**Abatement Costs**	**Environmental Degradation Costs**	**Ratio**	**Mean**
2004	922.3	2198	2.38	
2005	1610.9	2869	1.78	
2006	1821.5	3051	1.67	1.93
2007	2104.8	3616.7	1.71	
2008	2227.7	4725.6	2.12	
Water Pollution (Unit: Hundred Million Yuan)				
Year	**Abatement Costs**	**Environmental Degradation Costs**	**Ratio**	**Mean**
2004	1808.7	2862.8	1.58	
2005	2084	2484.7	1.19	
2006	2143.8	2705.8	1.26	1.33
2007	2121.1	2774.8	1.31	
2008	2672.6	3457.1	1.29	
Solid Waste Pollution (Unit: Hundred Million Yuan)				
Year	**Abatement Costs**	**Environmental Degradation Costs**	**Ratio**	**Mean**
2004	143.5	26.5	0.18	
2005	148.7	29.6	0.20	
2006	147.3	29.6	0.20	0.31
2007	129.8	65.1	0.50	
2008	142.9	63.6	0.45	

3. Case Study and Results

3.1. Case Description

To compare the environmental cost of residential buildings during the operational stage, two sites were selected as case study buildings. One is in Beijing, and the other is in Xiamen, which represent two different climate zones in China. According to the construction organization flow chart, the construction period is 2 years. The major construction materials include concrete, rebar, steel tube, cement mortar, wood, aluminum, glass and alkyd paint. The specific information and corresponding direct costs of the two case buildings are shown in Table 8.

Table 8. Comparison of the case study buildings in Beijing and Xiamen.

Case Study Building	Gross Floor Area	Climate Zone	Number of Apartments	Direct Costs
In Beijing	63,627 m^2	North temperate subhumid continental monsoon climate	100	266,784.3 thousand CNY
In Xiamen	12,595 m^2	Subtropical marine monsoon climate	78	39,794.76 thousand CNY

Data and specifications required for this study are obtained from the structural drawings of the buildings, acceptable LCI database and other archived literature. The materials used for construction are specified in the bill of quantities and can be obtained from the contractors.

3.2. Results Analysis

Based on the calculation method mentioned above, per capita energy consumption level of residents in Xiamen and Beijing were calculated based on China's Yearbook. The environmental costs of the two case studies are shown in Figures 2–5.

Most of the input data used in this case study comes from actual utility bills. However, due to the inevitable limitations of the input data, corresponding assumptions were made during the analysis. Considering the variability of critical input variables, sensitivity analysis of key parameters was conducted. Sensitivity analysis is the measurement of changes in one or more uncertainties to determine the extent to which changes in each factor affect the expected objective [34]. In this paper, the single-factor sensitivity analysis method is used to quantitatively describe the importance degree of input variables when only one parameter changes by 1%. The calculation formula is as shown in Equation (9).

$$E_i = \Delta C_{ei} / \Delta F_i \tag{9}$$

where E_i is the sensitivity parameter of the variable F_i; ΔC_{ei} is the corresponding rate of change in environmental costs (%); ΔF_i is the rate of change of the variable F_i, taken as 1%.

To find the critical input variables, the sensitivity analysis results are shown in Figure 2. For both of the case study buildings, the sensitivity coefficient of the unit virtual abatement cost of CO_2 is the largest, equaling 0.67, which means that CO_2 has the largest influence on the final environmental cost. Additionally, the unit virtual abatement cost of N_2O, CH_4 and NO_x are also key parameters that may lead to significant changes in the outcome, with values of 0.12, 0.45 and 0.32 respectively. The environmental cost results are not sensitive to the unit virtual abatement cost of CO, COD, dust, NH_4^+, SO_2, solid waste and VOC.

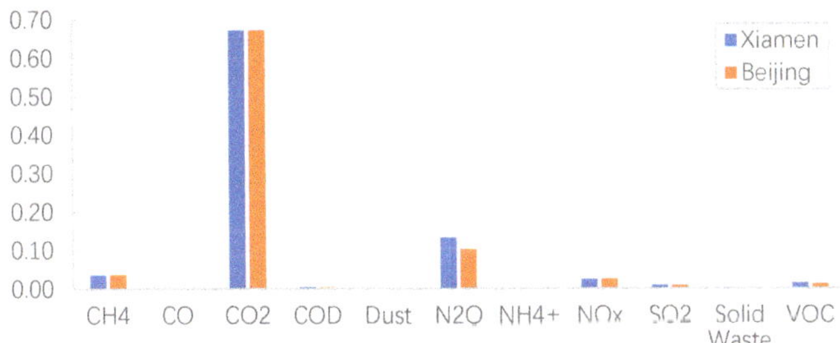

Figure 2. Sensitivity coefficients of the unit virtual abatement costs.

Additionally, the data quality indicators (DQI) method [35] (see Table 9) and Monte Carlo simulation were used in this case study to analyze the LCA data quality and uncertainty of the results. The engineering quantity data are all from the engineering quantity list, and the emission factor data are from a database. According to the standard deviation provided in Eco-invent [36], the distribution type of the LCI data is selected as lognormal distribution, and the uncertainty is shown in Table 10. Using the Monte Carlo simulation, the variability of environmental scores associated with the ratio of green construction measures cost to each subengineering fees, transportation distance and the average fuel consumption for each vehicle can be estimated. The selected variables are assumed to be uniform distribution or lognormal distribution (see Table 10), and 10,000 iterations were carried out based on previous studies [37].

Table 9. Data quality indicators (DQI) and uncertainty.

Indicator Score	1	2	3	4	5
Reliability	Verified data based on measurement (0.0)	Verified data party based on assumptions or nonverified data based on measurements (0.025)	Nonverified data partly based on qualified estimates (0.05)	Qualified estimate (e.g., by industrial expert) (0.1)	Nonqualified estimate (0.2)
Completeness	Representative data from all sites relevant to the market considered over an adequate period to balance normal fluctuations (0.0)	Representative data from >50% of the sites relevant to the market considered over an adequate period to balance normal fluctuations (0.01)	Representative data from only some sites (<<50%) relevant to the market considered or >50% of the sites but from shorter periods (0.025)	Representative data from only one site relevant to the market considered or some sites but from shorter periods (0.05)	Representativeness unknown or data from a small number of sites and from shorter periods (0.1)
Temporal correlation	Less than 3 years of time difference from the dataset (0.0)	Less than 6 years of time difference from the dataset (0.025)	Less than 10 years of time difference from the dataset (0.1)	Less than 15 years of time difference from the dataset (0.2)	Age of data unknown or more than 15 years of time difference from the dataset (0.35)
Geographical correlation	Data from area under study (0.0)	Average data from larger area in which the area under study is included (0.015)	Data from area with similar production conditions (0.05)	Data from area with slightly similar production conditions (0.1)	Data from unknown or distinctly different area (0.2)
Further technological correlation	Data from enterprises, processes and materials being studied (0.0)	Data from processes and materials being studied (i.e., identical technology) but from different enterprises (0.005)	Data from processes and materials being studied but from different technology (0.01)	Data on related processes or markets (0.025)	Data on related processes on laboratory scale or from different technology (0.05)

Table 10. Data uncertainty of each parameter in the calculation.

Parameter	Variability	Distribution	Iteration Times
Material quantity	0.035	Lognormal	10,000
Data quality indicator	0.0433	Lognormal	10,000
Transportation distance	20–40	Uniform	10,000
Average fuel consumption	10–15	Uniform	10,000
Ratio of green construction measures cost to construction fees	0.59%	Lognormal	10,000
Ratio of green construction measures cost to decoration fees	5.18%	Lognormal	10,000
Ratio of green construction measures cost to erection fees	12.25%	Lognormal	10,000

Figures 3 and 4 show that the added variability did not significantly change the average values nor did it change the ranking of the four stages in terms of C_{va}, C_{ed} and C_{gc}. The minimum, average and maximum total environmental costs are 412, 616 and 827 CNY/m^2, respectively, in the Xiamen case study building, while they are 489, 673 and 899 CNY/m^2, respectively, in the Beijing case study building. The coefficient of variation of the material production stage is the largest, followed by the operation and maintenance stage, while that of the demolition stage is the smallest.

The average value of the case study building in Xiamen is shown in Figure 3, where the total environmental cost is 616.29 CNY/m^2, of which the biggest contributor to environmental cost is material production stage reaching 330.96 CNY/m^2, followed by operation stage, 199.40 CNY/m^2. The environmental cost of demolition stage is negative, which indicates that the recycled material can bring positive environmental benefit. For the case study building in Beijing (shown in Figure 4), the total environmental cost is 672.80 CNY/m^2. The environmental cost of material production stage is 307.42 CNY/m^2, followed by operation stage, 247.07 CNY/m^2, which is slightly higher than that of the Xiamen case building's operation stage. This is possibly because energy consumption of heating is excluded for the case study building in Xiamen, which is located in a hot-summer and warm-winter zone where heating in the winter is not necessary. For the both case study buildings, construction stage is the third largest contributor to the environmental cost of the life cycle. During this stage, the green construction cost accounts for the largest percentage of the total environmental cost, about 65%. Demolition stage has the minimum environmental cost. For the both case study buildings, C_{ed} accounts for about 69% of the total environmental cost during material production stage, operation stage and demolition stage.

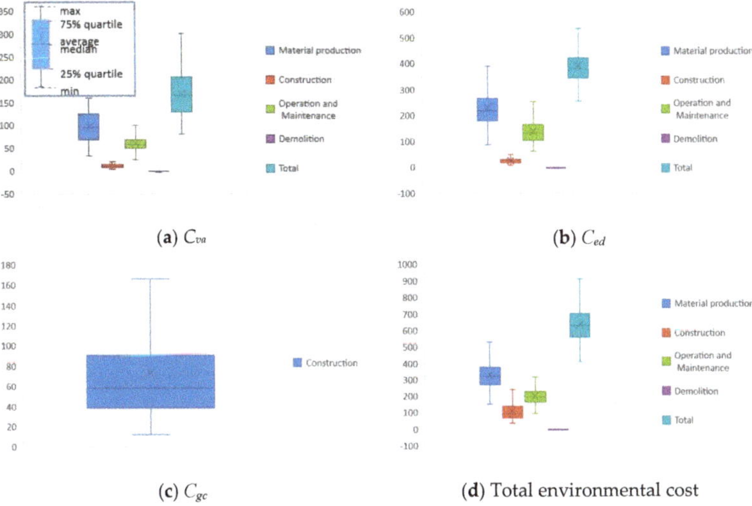

(**a**) C_{va}

(**b**) C_{ed}

(**c**) C_{gc}

(**d**) Total environmental cost

Figure 3. Environmental cost of case study building in Xiamen (unit: CNY/m^2).

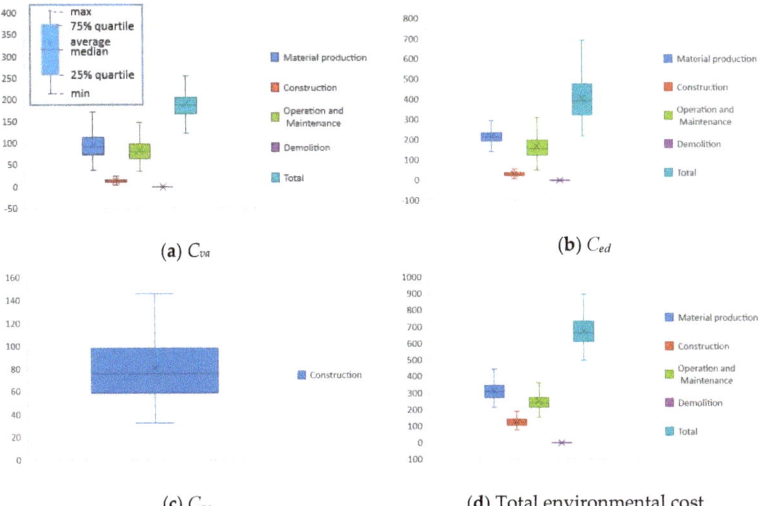

(**a**) C_{va}

(**b**) C_{ed}

(**c**) C_{gc}

(**d**) Total environmental cost

Figure 4. Environmental cost of case study building in Beijing (unit: CNY/m^2).

According to the results of both case studies (see Figure 5), a default discount rate was selected as 7%; the environmental cost in life cycle achieves an indispensable 14% share of the direct cost. However, the existing direct cost of the life cycle often neglects environmental cost, resulting in a great warp between calculation results and actual results. In some research, environmental costs are roughly assumed as 10% of direct costs. Since the percentage adopted is less than the result of this case study, this would lead to an error. In order to consider the variability of discount rate, therefore, this study assumes discount rates of 7%, 12% and 17%. The uncertainty analysis of the two cases shows that the ratio of environmental cost to direct life cycle cost decreases as the discount rate increases, but the change does not exceed 5%.

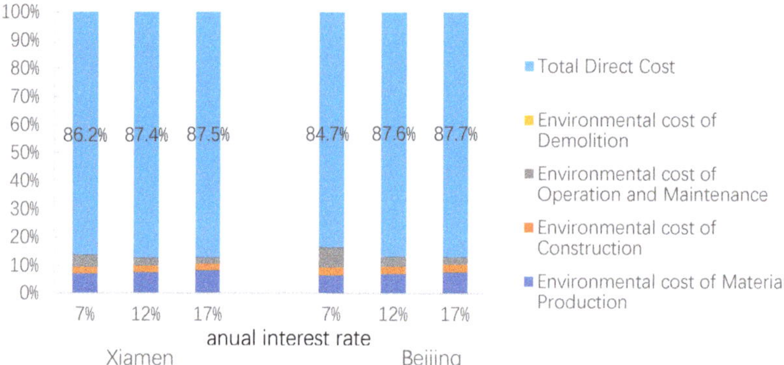

Figure 5. Ratio of environmental cost to direct cost considering different discount rate.

4. Conclusions and Limitations

4.1. Conclusions

Different from the previous LCCA studies which neglect or roughly estimate environmental cost, this study calculates environmental cost based on life cycle inventory. Besides, this study presents a basic method to improve the calculation of LCCA of a building during its life cycle. Furthermore, a single-objective optimization model is generated by converting environmental impact into environmental cost, which has the same unit as LCC and can help the decision makers to obtain a single optimum design solution. Finally, a quantitative analysis of case study of residential buildings in Xiamen and Beijing has been conducted. Based on the above research, the following conclusions can be drawn:

- The environmental costs of residential buildings in Beijing and Xiamen are 679 CNY/m^2 and 640 CNY/m^2 respectively, of which the biggest contributor is material production, followed by operation stage, construction stage and demolition stage.
- For both of the two case study buildings, the environmental degradation cost accounts for about 70% of the total environmental cost, and environmental cost accounts for about 16% of direct cost.
- The sensitivity analysis results show that the unit virtual abatement cost of CO_2 has the largest influence on the final environmental cost, followed by N_2O, CH_4 and NO_x. The environmental cost results are not sensitive to the unit virtual abatement cost of CO, COD, dust, NH_4^+, SO_2, solid waste and VOC.
- The coefficient of variation of the material production stage is the largest, followed by operation and maintenance stage, while the demolition stage is the most robust.
- The uncertainty analysis of the two cases shows that the ratio of the environmental cost to the direct life cycle cost decreases as the discount rate increases, but the change does not exceed 5% when the discount rate varies from 7% to 17%.

4.2. Limitations

- The total environmental cost of the case study building in Xiamen is 50 CNY/m^2 lower than that of case building in Beijing. However, due to the lack of actual operation data, energy consumption data is analyzed based on the reference of the local yearbook, which represents the general operational energy cost of all buildings, including public buildings and residential buildings.

Further research should focus on using specific operation data to increase the reliability and accuracy of the estimation results.

- The theoretical approach proposed in this study is based on the SEEA, which is an incomplete green GDP accounting method. For example, it does not consider the loss caused by ecological damage, groundwater pollution and soil pollution. Further research will refine the methodology based on a new ISO standard [38] proposed in 2019.

- Due to the lack of data on construction waste disposal in China, this paper uses the recycle rate of building materials from foreign data. Localization of data for construction waste disposal is still required for in-depth study to obtain accurate values.

Author Contributions: Y.W. provided the idea and led the research work. B.P., X.Z. and J.W. conceptualized the problem, performed the proposed methodology and the case study and prepared original draft. Y.L., C.S. and S.Z. reviewed and edited the original draft. All authors discussed the results and contributed to the final manuscript. All authors have read and agreed to the published version of the manuscript.

Funding: This research has been supported by the National Key Research and Development Plan of China (grant number 2016YFC0701806) and China Railway Construction Real Estate Group North Co. LTD (grant number C19L01450).

Conflicts of Interest: The authors declare no conflicts of interest.

Nomenclature

C_e	Total life cycle environmental costs
C_{gc}	Green construction measures costs
C_{va1}	Virtual abatement costs of air pollution
C_{va2}	Virtual abatement costs of water pollution
C_{va3}	Virtual abatement costs of solid waste pollution
C_{va31}	Virtual abatement costs of building solid waste
C_{va32}	Virtual abatement costs of household waste
C_{ed1}	Environmental degradation cost of air pollution
C_{ed2}	Environmental degradation cost of water pollution
C_{ed3}	Environmental degradation cost of waste pollution
C_{epv}	Present value of the total environmental costs
C_{ep}	Environmental cost of material production stage
C_{ec}	Environmental cost of construction stage
C_{eo}	Environmental cost of operation stage
C_{edem}	Environmental cost of demolition stage
Q	Diesel consumption
Q_1	Amount of air pollutants
Q_2	Amount of water pollutants
Q_{31}	Total amount of building solid waste
Q_{32}	Total amount of household waste
Q_k	Amount of household waste treated by different technologies
c_{va1}	Unit virtual abatement costs of air pollution
c_{va2}	Unit virtual abatement costs of water pollution
c_{va31}	Virtual abatement cost of per unit building solid waste
c_{va32}	Transportation costs of household waste
c_{vak}	Unit virtual abatement costs of each treatment
m_i	Mass of *i*-th material
L_i	Transportation distance of *i*-th material, assumed to be 181 km
q_{mi}	Average fuel consumption for transporting per unit material, assumed to be 101.78L/(kt·km)
r_i	Average ratio of environmental degradation costs to virtual abatement costs

t_1	Number of annual interest periods during construction stage, assumed to be 2 years
t_2	Number of annual interest periods during operation stage, assumed to be 50 years
r	Discount rate, assumed to be 7%
A	Equal annual payment
ΔC_{ei}	Corresponding rate of change in environmental costs (%)
ΔF_i	Rate of change of the variable F_i, taken as 1%
F_i	Sensitivity parameter of the variable F_i

References

1. Ministry of Housing and Urban-Rural Development of the People's Republic of China. *Statistical Bulletin on Urban and Rural Construction in 2014*; Ministry of Housing and Urban-rural Development: Beijing, China, 2015.

2. National Bureau of Statistics. *China Statistical Yearbook*; China Statistics Press: Beijing, China, 2014.

3. United States; Environmental Protection Agency; Office of Pollution Prevention; ICF Incorporated. *An Introduction to Environmental Accounting as a Business Management Tool: Key Concepts and Terms*; Office of Pollution Prevention and Toxics Press: New York, NY, USA, 1995.

4. United Kingdom (UK). *System of National Accounts*; China Statistics Press: Beijing, China, 1995.

5. Yu, F.; Wang, J.N.; Cao, D. *Guidelines for Chinese Environmental and Economic Accounting*; China Environmental Science Press: Beijing, China, 2009.

6. Yu, F.; Wang, J.N.; Cao, D. *Chinese Environmental and Economic Accounting Report 2004*; China Environmental Science Press: Beijing, China, 2009.

7. Yu, F.; Wang, J.N.; Cao, D. *Chinese Environmental and Economic Accounting Report 2007–2008*; China Environmental Science Press: Beijing, China, 2012.

8. Yu, Y.; Chen, D.; Zhu, B. Eco-efficiency trends in China, 1978–2010: Decoupling environmental pressure from economic growth. *Ecol. Indic.* **2013**, *24*, 177–184. [CrossRef]

9. Lehman, G. Global accountability and sustainability: Research prospects. *Account. Forum* **2002**, *26*, 219–232. [CrossRef]

10. Traverso, M.; Asdrubali, F.; Francia, A.; Finkbeiner, M. Towards life cycle sustainability assessment: An implementation to photovoltaic modules. *Int. J. Life Cycle Assess.* **2012**, *17*, 1068–1079. [CrossRef]

11. Kravanja, Z.; Čuček, L. Multi-objective optimization for generating sustainable solutions considering total effects on the environment. *Appl. Energy* **2013**, *101*, 67–80. [CrossRef]

12. Burnett, R.D.; Hansen, D.R. Ecoefficiency: Defining a role for environmental cost management. *Account. Org. Soc.* **2008**, *33*, 551–581. [CrossRef]

13. Vogtländer, J.; Van der Lugt, P.; Brezet, H. The sustainability of bamboo products for local and Western European applications. LCAs and land-use. *J. Clean. Prod.* **2010**, *18*, 1260–1269. [CrossRef]

14. Baeza-Brotons, F.; Garcés, P.; Payá, J.; Saval, J.M. Portland cement systems with addition of sewage sludge ash. application in concretes for the manufacture of blocks. *J. Clean. Prod.* **2014**, *82*, 112–124. [CrossRef]

15. Kendall, A.; Keoleian, G.A.; Helfand, G.E. Integrated life-cycle assessment and life-cycle cost analysis model for concrete bridge deck applications. *J. Infrastruct. Syst.* **2008**, *14*, 214–222. [CrossRef]

16. Chen, F.F. Life Cycle Environmental Costs of Bridge Engineering. Ph.D. Thesis, Wuhan University of Technology, Hubei, China, 2011.

17. Carreras, J.; Boer, D.; Cabeza, L.F.; Jiménez, L.; Guillén-Gosálbez, G. Eco-costs evaluation for the optimal design of buildings with lower environmental impact. *Energy Build.* **2016**, *119*, 189–199. [CrossRef]

18. Chou, J.S.; Yeh, K.C. Life cycle carbon dioxide emissions simulation and environmental cost analysis for building construction. *J. Clean. Prod.* **2015**, *101*, 137–147. [CrossRef]

19. Epstein, M.; Flamholtz, E.; McDonough, J. Corporate social accounting in the United States of America: State of the art and future prospects. *Account. Organ. Soc.* **1976**, *1*, 23–42. [CrossRef]

20. International Organization for Standardization, 14044. *Environmental Management–Life Cycle Assessment—Requirements and Guidelines*; I.S.O.: Geneva, Switzerland, 2006.

21. Society of Environmental Toxicology and Chemistry. *Guidelines for Life-Cycle Assessment: A "Code of Practice"*; SETAC: Brussels, Belgium, 1993.

22. Surahyo, M.A. Environmental and social cost modeling of highway projects. Ph.D. Thesis, University of Toronto, Toronto, ON, Canada, 2005.
23. European Reference Life Cycle Database (ELCD). Available online: http://lca.jrc.ec.europa.eu/lcainfohub/datasetCategories.vm (accessed on 12 May 2012).
24. Beijing Municipal Commission of Housing and Urban-Rural Development (BMCHUD). *Quota of Beijing Construction Project*; China Construction Industry Press: Beijing, China, 2012.
25. Ministry of Construction of the People's Republic of China. *National Unified Construction Machinery Quota*; China Planning Press: Beijing, China, 2014.
26. Beijing Statistics Bureau. *Yearbook of Beijing*; Chinese Statistic Press: Beijing, China, 2016.
27. Xiamen Statistics Bureau. *Yearbook of Xiamen*; Chinese Statistic Press: Xiamen, China, 2016.
28. The Low-Carbon Buildings Method 3.0. Available online: www.lcbmethod.com (accessed on 14 May 2017).
29. Zeng, H. Life cycle assessment of environmental impacts on construction waste disposal. *Eco-Economic* **2013**, *23*, 132–135.
30. Lei, Y.J.; Yao, J.; Zhao, F.; Li, Z.S. Environmental impact assessment of two kinds of treatment methods of municipal solid waste in Chengdu. *Saf. Environ. Eng.* **2014**, *21*, 75–79.
31. Xiao, J.Z. *Recycled Concrete*; China Building Industry Press: Beijing, China, 2008.
32. Zhu, H.; Xiu, L.I.; Liu, L.; Wang, C. The harmless disposal of urban domestic waste and sustainable development. *Environ. Sci. Technol.* **2002**, *25*, 22–26.
33. Thuesen, G.J.; Fabrycky, W.J. *Engineering Economy*; Pearson Education Asia Limited and Tsinghua University Press: Beijing, China, 2004.
34. Ardente, F.; Beccali, G.; Cellura, M.; Brano, V.L. Life cycle assessment of a solar thermal collector: Sensitivity analysis, energy and environmental balances. *Renew. Energy* **2005**, *30*, 109–130. [CrossRef]
35. Kennedy, D. The Diet Quality Index-International (DQI-I) provides an effective tool for cross-national comparison of diet quality as illustrated by China and the United States. *J. Nutr.* **2003**, *133*, 3476–3484.
36. Weidema, B.P.; Bauer, C.; Hischier, R.; Mutel, C.; Nemecek, T.; Reinhard, J.; Vadenbo, C.O.; Wernet, G. *Overview and Methodology: Data Quality Guideline for the Ecoinvent Database Version 3*; Swiss Centre for Life Cycle Inventories: Swiss, Switzerland, 2013.
37. Tatari, O.; Nazzal, M.; Kucukvar, M. Comparative sustainability assessment of warm-mix asphalts: A thermodynamic based hybrid life cycle analysis. *Resour. Conserv. Recycl.* **2012**, *58*, 18–24. [CrossRef]
38. International Organization for Standardization, 14008. *Monetary Valuation of Environmental Impacts and Related Environmental Aspects*; I.S.O.: Geneva, Switzerland, 2019.

© 2020 by the authors. Licensee MDPI, Basel, Switzerland. This article is an open access article distributed under the terms and conditions of the Creative Commons Attribution (CC BY) license (http://creativecommons.org/licenses/by/4.0/).

Article

The Impacts of a Building's Thermal Mass on the Cooling Load of a Radiant System under Various Typical Climates

Rong Hu [1], Gang Liu [2,*] and Jianlei Niu [3]

[1]   School of Architecture and Traffic Engineering, Guilin University of Electronic Technology, Guangxi 541004, China; rong.hu@connect.polyu.hk
[2]   School of Energy Science and Engineering, Central South University, Changsha 410083, China
[3]   Department of Building Services Engineering, The Hong Kong Polytechnic University, Hong Kong 999077, China; jian-lei.niu@polyu.edu.hk
*   Correspondence: gangliu@csu.edu.cn

Received: 20 February 2020; Accepted: 9 March 2020; Published: 14 March 2020

Abstract: Cooling load is difficult to predict for a radiant system, because the interaction between a building's thermal mass and radiation heat gain has not been well defined in a zone with a cooling surface. This study aims to reveal the effect of thermal mass in an external wall on the transmission load in a space with an active cooling surface. We investigated the thermal performances in a typical office building under various weather conditions by dynamic simulation with Energy-Plus. It was found that the thermal mass in the inside concrete layer had positives in terms of indoor temperature performance and energy conservation. The peak cooling load of the hydronic system decreases 28% in the proper operating state, taking into account the effect of the thermal mass in an external wall. Compared to the performances in zones with equivalent convective air systems (CASs), the peak cooling load and the accumulated load of the combined system (radiant system coupled by fresh air system) are higher by 9%–11% and 3%–4%, respectively. The effect of thermal mass is evident in a transient season with mild weather, when the relative effects are about 45% and 60%, respectively, for a building with radiant systems and a building with equivalent CASs.

Keywords: thermal mass; thermal inertia; radiant cooling system; energy conservation; energy simulation

---

1. Introduction

The energy consumption of space heating and cooling has attracted attention, and thermal transmittance (U-value) must be limited to maximum acceptable values for commercial and residential buildings according to construction regulations and related energy efficiency standards [1–3]. Therefore, increasing the normal thermal resistance (R-value in steady state) of a building envelope is the main measure used to protect the indoor environment from extreme external conditions and reduce the energy consumption on space heating and cooling. However, these factors are not sufficient to characterize the dynamic thermal behavior. In a transient situation, the thermal mass of a structure can store or release heat depending on the surrounding temperature differences. Balaras [4] reviewed tools for calculating cooling load, accounting for thermal mass, and indicated the effectiveness of thermal mass on an indoor thermal environment and energy conservation, particularly in the places with deep diurnal temperature differences. Regulating the amount of thermal mass can increase the time lag and decrease the temperature fluctuation in a conditioned space. Optimizing the thermal mass has been regarded as an important measure for passive heating/cooling strategies and for designing low-energy buildings [5–7]. Besides, the phase change materials (PCMs) embedded in a building enclosure are regarded as a useful passive method to increase in heat storage capacity and thermal inertia further [8]. The structures are expected to narrow indoor temperature fluctuations and reduce energy demands [9], and the integration of PCM technologies has been on trial in some net-zero energy buildings recently [10,11].

Many studies have focused on the investigation of thermal behavior for an individual building envelope with the objective to optimize arrangement of insulation and massive layers. Al-Sanea [12] developed a concept of dynamic thermal resistance, accounting for the influences of wall orientation, long wave radiation exchange, thermal energy storage, and nominal thermal resistance. With this concept, Al-Sanea et al. [13,14] investigated the effects of insulation locations and various amounts of thermal mass on thermal performances of building external walls based on the climate of Riyadh. They recommended that building walls should contain a minimum critical amount of thermal mass, and that the insulation be placed on the outside in a case where the air conditioning system runs continuously; otherwise the insulation layer should be placed on the inside. Tsilingiris [15–17] investigated the effects of various insulation configurations and heating systems on the energy loss through a building envelope. The results showed that the position of a massive layer strongly influences the transient heat transfer through the structure, but that it has no effect on the heat flux in the time-average quasi steady-state; and they also confirmed that the thermal insulation performs better when located at the inside in an intermittently conditioned room. Deng et al. [18] suggested that the high thermal mass in an external wall should be directly faced toward the indoor air to avoid overheating in the part-time, part-space cooling conditions. An experimental study by Kumar et al. [19] showed that a high inside mass can help to reduce excess heat discomfort for a significant time; i.e., 40% and 98% of the summer and winter respectively in a naturally ventilated office building in India. Reilly and Kinnane [20] developed new metrics (transient energy ratio and effective U-value) to quantify the effects of thermal mass on the energy use for heating and cooling. They found that a high thermal mass possibly causes reductions in energy use in a hot climate with big diurnal temperature differences, but it could lead to more energy use in cold climates. In addition, with the building energy simulation tools being well-developed, the transient thermal behavior can be investigated, and the effect of thermal mass on energy performance can be evaluated in a building approximating real-life, wherein the occupancy gain, solar gain, and HVAC operation strategies can be involved. Rodriguez et al. [21] stated that human behavior is one of the most important factors when understanding building physics. Eben Saleh [22,23] used a computer program named National Bureau of Standards Load Determination to analyze the energy use of an entire building and showed that placing the insulation on the outside of the building envelope can improve performance. Kossecka and Kosny [24,25] utilized simulation software DOE to investigate the thermal performances of six different configurations used in residential buildings in different USA climates. They recommended that a wall with an internal insulation layer can improve performance in a continuously used residential building, but indicated that a wall with inside insulation can enhance performance for intermittent heating and cooling. Verbeke and Audenaert [26] reviewed the impacts of thermal inertia in buildings across climate and building use, and suggested that assessing the impact of thermal inertia should be based on studies on the scale of whole buildings. According to their conclusion, the impacts of thermal inertia on energy use are relatively small and variable, with both positive and negative performances existing, but the thermal inertia can be used to shift the peak-load of an HVAC system in a proper control strategy.

In the light of the previous studies, additional problems have still not been focused on or clearly addressed:

1. Previous studies were based on investigations of indoor air temperature, and few considered envelope surface temperatures;
2. Many studies were based on situations wherein conventional heating and cooling systems were applied, without noting the differences in the heat transfer process for a radiant system.

At present, the radiant system has been widely used in recent years and is regarded to have many advantages in indoor environmental control and energy efficiency [27]. As an alternative cooling method, the radiant system is suggested to make use of the construction thermal mass to shift the peak HVAC system cooling load and power use [26,28]. However, as stated by Niu et al. [29], the radiant effect of a chilled ceiling can decrease the heat storage capacity of a building envelope. Feng et al. [30,31] also indicated the cooling load differences between radiant and air systems through simulation and measurement verification. Several methods of operating the radiant system in practice are available [32,33], and appropriate scheduling techniques can provide some opportunities to reduce the energy consumption with smaller initial investments [34]. The present authors [35] have conducted research on the operating characteristics of two radiant systems using various strategies in a typical office building through simulative and on-site measurements. The thermal mass of a slab can be utilized for cooling storage to shift the peak cooling load, particularly for a thermally active building system (TABS) in which a hydronic system is deeply embedded in the slab. In addition, an experimental study by Tahersima et al. [36] showed that the mass in the radiant floor can also be used for heating conservation during off-peak hours, and the operational costs result in significant savings.

On the basis of those considerations, the present study analyzed the effects of thermal mass in external walls on transmission loads in spaces with radiant cooling systems, and we present the differences from an identical room equipped with only an equivalent convective air system (CAS). In addition, operative temperature was used to evaluate the thermal comfort level in a room with radiant surfaces [37–40]. That parameter combines room air temperature and radiant temperature, weighted by convection and radiation heat transfer coefficients respectively.

2. Methodology

In a zone with a CAS alone, convection heat gain directly becomes the cooling load of the CAS, whereas instantaneous radiation gain is absorbed and stored in the thermal mass, and then converted to be the cooling load by convection with a time lag (Figure 1a). In a zone with a radiant system (Figure 2), which usually acts as a cooling ceiling or a floor, a CAS as an auxiliary system is necessary to maintain the indoor hygiene level, and is responsible for the zone latent load. Thus, besides the convection gain which is immediately handled by the auxiliary CAS, a part of the radiation heat gain is directly absorbed by the cooling surface (active surface). The remaining radiation heat gain is absorbed by the structure or furniture and then extracted by these two systems simultaneously (Figure 1b).

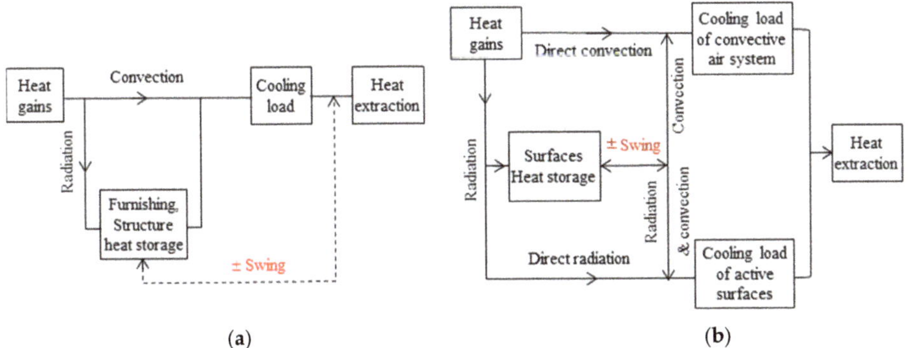

Figure 1. Cooling load generation schemes for (**a**) a convective air system [41] and (**b**) for a combined system (fresh air system + radiant cooling system) [35].

Dynamic simulation is generally a reliable method used to compute heat transfer in a given zone [42]. Feng et al. [30] stated that the heat balance method should be employed to calculate the cooling loads of radiant systems; and used the Energy-Plus simulation software to assess the cooling load. Energy-Plus was developed by U.S. Department of Energy and Lawrence Berkeley National Laboratory. The present algorithm models have been validated according to the standard method of test for the evaluation of building energy analysis computer programs (ANSI/ASHRAE 140) [43]. Thus, the Energy-Plus simulation software was also selected for computing instantaneous heat transfer in a building with cooling surfaces in this study. The algorithm is based on the conduction transfer function.

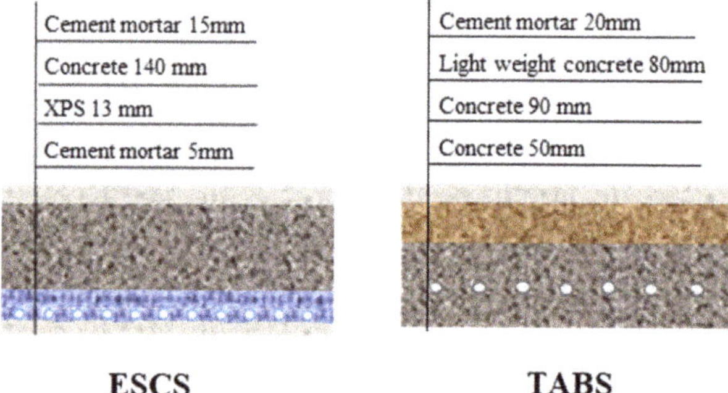

Figure 2. Configurations of radiant system.

In this study, the results of instantaneous heat extracted from a given zone (q_{ZH}), heat extracted by a cooling surface (q_{cs}), and the portion due to radiation (q_{ZRH}, q_{CSRH}) are significant for analysis, but they cannot be directly obtained from the simulation, and the deduction process is as follows.

In a conditioned space, heat balances of an inside face and the indoor air can be expressed as follows:

$$q''_{LWX} + q''_{SW} + q''_{LWS} + q''_{ki} + q''_{sol} + q''_{conv} = 0 \tag{1}$$

$$q_{conv} + q_{CE} + q_{IV} + q_{airsys} = 0 \tag{2}$$

Thus, total convection heat transfer from surfaces in an envelope can be expressed as follows:

$$\begin{aligned} q_{conv} = \sum_{i=1}^{surfaces} q''_{conv} = & -(\sum_{i=1}^{surfaces} q''_{LWX} + \sum_{i=1}^{surfaces} q''_{SW} + \\ & \sum_{i=1}^{surfaces} q''_{LWS} + \sum_{i=1}^{surfaces} q''_{ki} + \sum_{i=1}^{surfaces} q''_{sol}) \\ = & -(q_{CE} + q_{IV} + q_{air\ sys}) \end{aligned} \tag{3}$$

and the cooling load of the air-conditioning system can be expressed as follows:

$$\begin{aligned} q_{air\ sys} = & \sum_{i=1}^{surfaces} q''_{LWX} + \sum_{i=1}^{surfaces} q''_{SW} \\ & + \sum_{i=1}^{surfaces} q''_{LWS} + \sum_{i=1}^{surfaces} q''_{ki} + \sum_{i=1}^{surfaces} q''_{sol} - q_{CE} - q_{IV} \end{aligned} \tag{4}$$

In a zone where a hydronic system is contained in a concrete floor or ceiling (Figure 2), the heat can be conducted from both sides to the internal source. Thus, the heat extraction at cooling surfaces can be expressed in Equation (5) as follows.

$$q_{cs} = q_f + q_c = -[(\sum_{i=ceiling}^{floor} q''_{LWX} + \sum_{i=ceiling}^{floor} q''_{SW} + \sum_{i=ceiling}^{floor} q''_{LWS} + \sum_{i=ceiling}^{floor} q''_{sol}) + \sum_{i=ceiling}^{floor} q''_{conv}] \tag{5}$$

Thus, the heat extraction from the thermal zone (q_{ZH}) can be expressed as follows:

$$
\begin{aligned}
q_{ZH} = q_f + q_c + q_{air\,sys} &= \sum_{i=1}^{surfaces} q''_{LWX} + \sum_{i=1}^{surfaces} q''_{SW} + \sum_{j=1}^{surfaces} q''_{LWS} + \sum_{i=1}^{surfaces} q''_{ki} + \\
\sum_{i=1}^{surfaces} q''_{sol} - q_{CE} - q_{IV} + \left(q_f + q_c\right) &= \left(\sum_{i=1}^{surfaces} q''_{LWX} + \sum_{i=1}^{surfaces} q''_{SW} + \sum_{i=1}^{surfaces} q''_{LWS} + \right. \\
\left. \sum_{i=1}^{surfaces} q''_{sol}\right) &= -\left(\sum_{i=ceiling}^{floor} q''_{LWX} + \sum_{i=ceiling}^{floor} q''_{SW} + \sum_{i=ceiling}^{floor} q''_{LWS} + \sum_{i=ceiling}^{floor} q''_{sol}\right) + \\
\sum_{i=1}^{surfaces} q''_{ki} &- \left(\sum_{i=ceiling}^{floor} q''_{conv} + q_{CE} + q_{IV}\right)
\end{aligned}
\tag{6}
$$

However, the size of a radiant system cannot be directly assessed by the Energy-Plus simulation software, because only CAS is assumed when sizing a calculation. For the consideration of differences in the heat transfer process between the zones with and without a radiant system, hydronic systems are assumed in an initial simulation, and the parameters (pipe dimension, water flow rate, water inlet temperature, etc.) refer to many practical items. Thus, repeat computations must be implemented until the room operative temperature can meet the design criteria.

Instantaneous zone radiation heat gain q_{ZRH} is distributed to the surfaces according to their surface temperatures and shape factors; it is the sum of short wave and long wave radiation gains:

$$
q_{ZRH} = \sum_{=1}^{surfaces} q''_{LWX} + \sum_{i=1}^{surfaces} q''_{SW} + \sum_{i=1}^{surfaces} q''_{LWS} + \sum_{i=1}^{surfaces} q''_{sol}
\tag{7}
$$

Radiation heat gain at the active surfaces q_{CSRH} can be obtained using Equation (8):

$$
q_{CSRH} = \sum_{i=ceiling}^{floor} q''_{LWX} + \sum_{i=ceiling}^{floor} q''_{SW} + \sum_{i=ceiling}^{floor} q''_{LWS} + \sum_{i=ceiling}^{floor} q''_{sol}
\tag{8}
$$

When $q_{ZRH} \geq q_{CSRH}$, a part or all of the radiation heat gain is absorbed by the active surfaces via direct and indirect radiation transfer; for $q_{ZRH} < q_{CSRH}$, not only is the zone radiation heat gain absorbed, but also more conductive heat transfer on inactive surfaces is compensated by the active surfaces through radiation heat transfer. (Heat gain on the fenestration surface, excluding short-wave transmitted heat gain, is grouped under conduction heat gain for simplicity in this study.)

2.1. Influencing Parameters

The parameters that influence thermal mass performance generally include the thermal environment condition, construction, occupant scheduling, and the heating ventilation and air-conditioning (HVAC) system and its own operational strategy [4].

2.1.1. Thermal Environment Conditions

The effect of thermal mass is clear in the places with moderate climate or deep diurnal temperature differences [4,10,16]. This study selects two typical places in China as examples; namely, Beijing and Nanjing. According to the Chinese Building Climate Demarcation [44], Beijing is in the cold area, where the maximum air dry-bulb temperature reaches 34.73 °C on a cooling design day (e.g., July 21, and referenced as BJ_7/21) and varies over a large range (i.e., 8.6 K); meanwhile, Nanjing is in the hot summer and cold winter area, where the peak value is 35.1 °C on the cooling design day, and the temperature difference between the maximum and minimum values is 6.5 K. On a typical day in the transient season (e.g., June 14), the peak outdoor air dry-bulb temperature in Beijing remains high (i.e., 32.71 °C), but the minimum temperature decreases to 19.17 °C. By contrast, on the typical day in Nanjing (referenced as NJ_6/14), the temperature decreases, ranging from 19.25 °C to 27.89 °C. In addition, indoor design temperature can be set as 26 °C (operative temperature) during the occupied period (07:00–19:00).

2.1.2. Building and Construction

The simulation is based on a typical office building with 20 floors. The plan of a standard floor is illustrated in Figure 3a. The external window area accounts for 50% of the wall, and the exterior blinds work when the incident solar intensity exceeds 50 W/m². The external wall is mainly composed of extrusion polystyrene insulation (XPS) and concrete (Figure 3b), and the thermal properties of the structures meet the national building efficiency standard [3]. The internal heat gains and their scheduling are illustrated in Figure 3c.

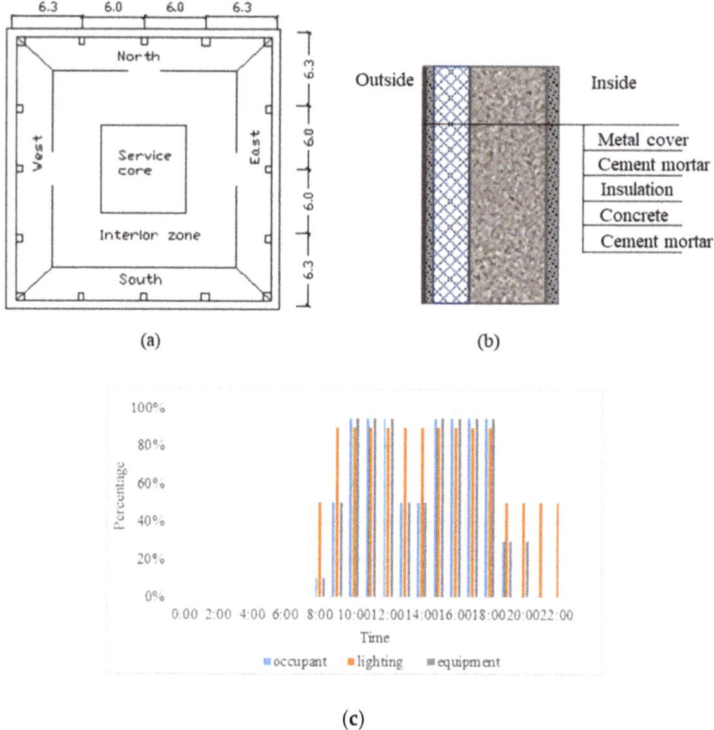

Figure 3. Building information. (**a**) The plan of a standard floor, (**b**) external wall construction (from outside to inside), (**c**) scheduling of internal heat gains in the building (heat gain from occupants: 12 W/m²; heat gain from lighting: 13 W/m²; heat gain from equipment: 20 W/m²).

The insulation and concrete in the external wall are sandwiched by two thin cement mortar layers, and the concrete layer is placed inside (Figure 3b). Thermal mass can be defined as the specific capacity multiplied by the mass of construction. For the 1D structure, the thermal mass per square meter of the external wall can be written as follows:

$$Thermal\ mass = C_{pi} \times \rho \times D \tag{9}$$

where C_{pi} is the specific capacity, J/kg·K; ρ is the density, kg/m³; and D is the layer thickness, m.

Given that specific capacity and material density are fixed, the layer thickness is the only function of the thermal mass of the construction. The normal thermal resistance of the external wall remains constant to distinguish the effects of thermal mass on the indoor environment, transmission load, and room sensible cooling load in a perimeter zone. Thus, the thermal mass increases as the thickness of the concrete layer increases, whereas the thickness of the insulation layer decreases. Table 1 shows that the thermal mass in external walls without concrete layer (lightweight structure) only accounts for 10% of the total in a perimeter zone on one standard floor in the office building, and the percentage increases to 43% as the thickness of the concrete layer extends to 200 mm.

Table 1. Concrete thickness of external opaque wall and the thermal mass in a perimeter zone on one standard floor in the office building.

-	Concrete Thickness	Insulation Thickness	Thermal Resistance of External Wall	Thermal Mass of External Wall	Percentage of the Total Thermal Mass in a Perimeter Zone	-
	mm	mm	m^2K/W	KJ/K·m^2	%	-
1	0	70.4	1.7	83.5	10%	light weight (LW)
2	50	69.0	1.7	198.4	22%	medium weight (MW)
3	100	67.7	1.7	313.4	31%	medium weight (MW)
4	200	65	1.7	543.2	43%	
5	300	62.3	1.7	773.1	52%	heavy weight (HW)
6	400	59.6	1.7	1003	59%	

2.1.3. HVAC System and Operational Strategy

The output of a radiant system should vary with climatic change, and the hydronic system has variable water flow rate but constant supply temperature. For considering the effect of thermal mass surrounding a hydronic system (the concrete slab acts significantly as a regenerator in TABS [35]), the embedded surface cooling system (ESCS) is only selected to determine the significant effect of thermal mass in the external walls on the heat transfer and transmission load in the zone with a radiant system.

The auxiliary CAS in the combined system (ESCS+CAS) is only responsible for fresh air cooling load and indoor latent load. It supplies conditioned air with a constant flow rate (1 ac/h) at a constant dry-bulb temperature (15 °C). Thus, the CAS is simulated with priority.

System operation strategy is a main factor which affects the thermal mass performance [13]. The conventional operation scheduling technique (OPCT), wherein the cooling system is available 24 h/day and the room temperature set point is adjusted to achieve energy savings, has been widely applied [45–48]. Thus, the auxiliary CAS runs continuously during the occupied period, and the hydronic system in ESCS operates according to the setting of the room thermostat.

An equivalent CAS is also considered in the identical perimeter zones and operates according to the strategy OPCT.

2.2. Definition

The parameters decrement factor and thermal phase lag, are frequently used to describe the effects of thermal mass in studies, and the specific definitions are listed as follows.

2.2.1. Decrement factor (f)

The decrement factor is a dimensionless factor which describes the change in amplitude of a sinusoidal heat transfer through a building component (Figure 4a), and it is generally expressed by the ratio of amplitude of temperature excitation to the value of response (Equation (10)) [26].

$$\text{decrement factor} = \frac{A_i}{A_e} = \frac{T_{i,max} - T_a}{T_{e,max} - T_a} \tag{10}$$

2.2.2. Thermal Phase Lag (φ)

The thermal phase lag determines how long it takes for excitation heat to go through an opaque material, and it represents the time difference between the moments when the maximum excitation and response heat fluxes occur (Figure 4b) [26].

According to the above definition, the instantaneous heat fluxes occurring on both sides of an opaque wall can be used to calculate decrement factor and thermal phase lag directly. The decrement factor in the article is defined by the ratio of the maximum conduction heat gain on the inside face to the maximum excitation heat on the outside face of an external wall.

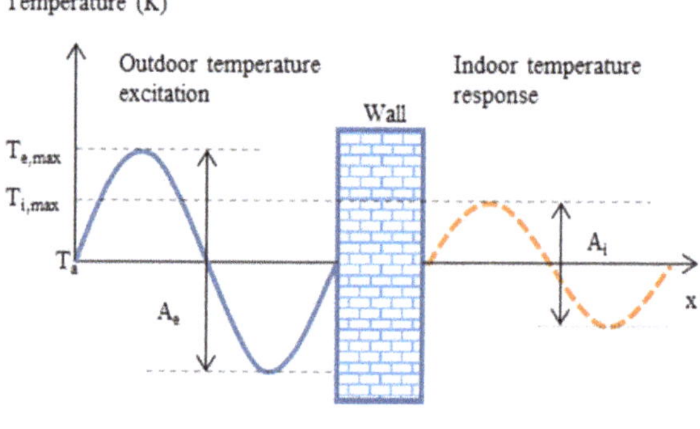

(**a**) Decrement factor.

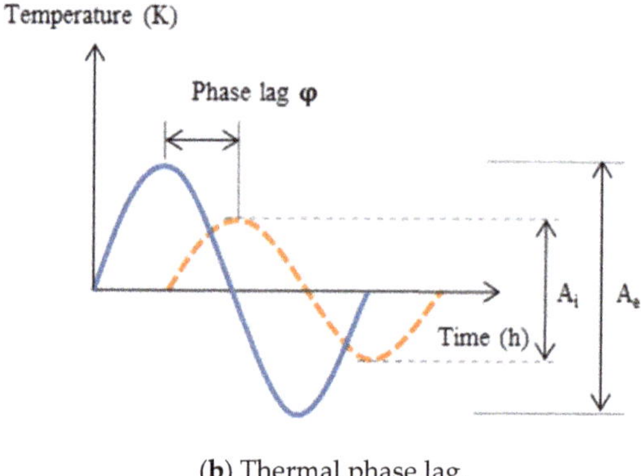

(**b**) Thermal phase lag.

Figure 4. Thermal response on a sinusoidal heat transfer through a wall [26].

2.2.3. Relative Effect (R)

In order to distinguish the effect of thermal mass in the zone with a combined system from the performance in the zone with an equivalent CAS, a parameter named relative effect is used and is defined as

$$Relative\ effect = \frac{f_{HW} - f_{LW}}{f_{LW}} \times 100\% \tag{11}$$

where $f_{H(L)W}$ is the decrement factor of a wall which is heavy (light).

3. Results

The effect of thermal mass on the room cooling load is limited in the perimeter zones. The peak room sensible cooling load and the accumulated load ($\sum_{24h} q_{ZT}$) decrease by 1% to 2% as the thickness of the concrete layer in the external wall increases from 0 to 200 mm. The data shows only a slight change as the thermal mass increases further (Figure 5). There are two main reasons. The first is that the conduction heat gain only accounts for a small portion of the total heat gain in the building at the peak time when the maximum building cooling load occurs; that is, only 3% or less in the perimeter zones

(Figure 6). The second reason can be illustrated by Figure 7; only a very narrow gap in accumulated transmission load exists between the structures of light weight and of heavy weight when the same cooling system applied to maintain an identical indoor thermal environment. It also confirms some preceding research: the thermal mass has little impact on building energy consumption [15–17,27].

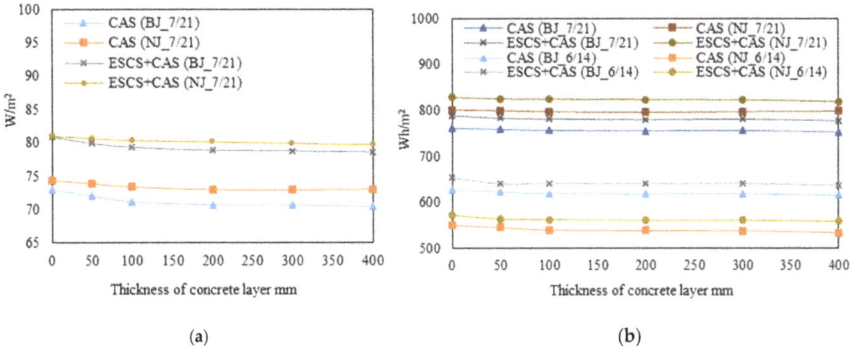

(a)   (b)

Figure 5. Effects of thermal mass on cooling loads of the combined system. (**a**) Effects of thermal mass on peak cooling loads on design days. (**b**) Effects of thermal mass on accumulated sensible cooling loads on typical days.

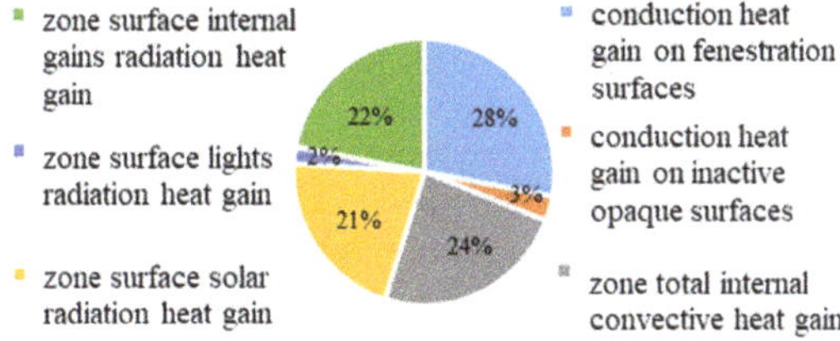

Figure 6. Composition of accumulated cooling load in perimeter zones with lightweight construction on a Nanjing cooling design day (embedded surface cooling system (ESCS)).

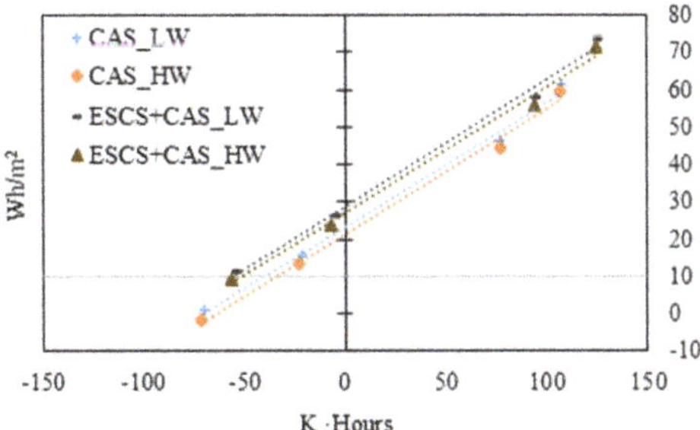

Figure 7. Accumulated conduction heat transfer rate through an external wall against degree-hours.

In addition, the peak cooling load of the zone with an ESCS is higher than the load of the zone with an equivalent CAS by 9% to 11%, and the accumulated load is also higher by 3%–4% based on same weather conditions. As Figure 7 illustrates, accumulated conduction heat gain through an external wall with a heavy thermal mass in the zone with an ESCS is still more than the amount on the wall with a lightweight structure in the zone with an equivalent CAS when the degree-hours are selfsame. These results are basically consistent with the findings from the laboratory experiment conducted by Woolley et al. [49]. The tests were carried out in a standard climatic chamber, and two separated tests were implemented side-by-side: one with a radiant cooling ceiling, and the other with an overhead mixing air distribution system. The results showed that the peak cooling load of the radiant system was 2%–10% larger than the load of air system in the case where internal heat gain varied periodically, and the accumulated load was 2%–7% higher. The experiment also confirmed the differences existing in heat transfer processes between the zones with and without a radiant system, but it did not explain the effect of thermal mass further.

The instantaneous transmission load has obvious varietious between the constructions with different thermal masses. As Tables 2 and 3 state, the internal thermal mass helps to decrease the transmission through external structures in different orientations, especially on the typical days in the transient season. By comparing the performances in the zone with an equivalent CAS, the effect of thermal mass is more evident in the zone with an ESCS on the cooling design days, while it becomes less on the days in transient season.

In addition, the internal thermal mass prolongs the phase lag for some structures with various orientations, especially on the typical days in transient season (Tables 2 and 3).

According to the impact factors mentioned above, the climatic situation, and the thermal mass position, the specific analysis is as follows by taking the performance in the north perimeter zone as an example.

3.1. Effect of Thermal Mass on System Performance in Different Climates

3.1.1. Performance on Cooling Design Days

On the cooling design days in the cities of Nanjing and Beijing, the minimum outdoor dry-bulb temperatures are higher than the room setting temperature; i.e., 26 °C. The inside face temperatures of external opaque walls are lower than the temperatures of the outside faces during the occupied period, for the structures with and without a concrete layer. In addition, for the external opaque surfaces which contain the same structure, the inside face temperatures appear lower in the zones with ESCSs compared to the temperatures in the zones with CASs (Figure 8a). Since the heat transfer process in the zone with an ESCS is different from the process in the zone with a CAS (Figure 1), most of instantaneous radiant heat gain can be absorbed by the cooling surface through direct or indirect radiant heat transfer (Figure 9a). This also confirms the statement by Niu et al. [29]: the cooling surface can decrease the heat storage capacity of the building envelope to the radiation heat transfer. Besides, a big portion of conduction occurring on the inside faces of the external walls is balanced by radiation instead of convection (averages of about 66% and 88%, respectively, for the zone with heavy weight and the zone with light weight). The instantaneous radiation heat fluxes on the inside faces are related to the cooling surfaces during the occupied period (Figure 9a), such that heat fluxes are conducted from the outside faces to the inside in this time, and the instantaneous conduction heat gains on the inside faces have approximate values in the most of the occupied time when these cooling systems operate to maintain the indoor environment within the given criteria (Figure 10a). However, the thermal mass in the external wall can help to maintain the inside face temperature stably, and the transmission loads through the external walls with heavy weights are lesser than the ones through the walls without concrete layers (Figure 10a). Consequently, compared to the performance in the zone with an equivalent CAS, the maximum conduction heat transfer on the inside face is slightly lower in the zone with an ESCS, and the relative effect of thermal mass (R) appears more significant on the cooling design days.

Table 2. The effects of thermal mass on thermal transmission (cooling design days).

| Conduction Heat Transfer | | Beijing(7–21) | | | | | Nanjing(7–21) | | | | |
| | | Decrement Factor | | Relative Effect | Phase Lag | | Decrement Factor | | Relative Effect | Phase Lag | |
Thermal Mass Weight		LW	HW		LW	HW	LW	HW		LW	HW
South	CAS	0.16	0.10	−40%	6	9	0.18	0.13	−26%	5	8
	ESCS	0.17	0.08	−53%	6	9	0.19	0.11	−42%	6	9
North	CAS	0.12	0.08	−33%	12	12	0.13	0.10	−25%	12	12
	ESCS	0.12	0.07	−43%	12	12	0.13	0.08	−35%	12	12
East	CAS	0.09	0.06	−33%	5	12	0.09	0.06	−33%	5	12
	ESCS	0.09	0.05	−49%	5	12	0.10	0.05	−45%	5	12
West	CAS	0.21	0.08	−65%	4	4	0.22	0.10	−56%	5	5
	ESCS	0.21	0.06	−70%	4	4	0.22	0.08	−66%	5	5
Building	CAS	0.16	0.11	−31%	12	12	0.15	0.12	−22%	12	12
	ESCS	0.15	0.09	−40%	12	12	0.15	0.10	−34%	12	12

Table 3. The effects of thermal mass on thermal transmission (typical days in transient season).

| Conduction Heat Transfer | | Beijing (6–14) | | | | | Nanjing (6–14) | | | | |
| | | Decrement Factor | | Relative Effect | Phase Lag | | Decrement Factor | | Relative Effect | Phase Lag | |
Thermal Mass Weight		LW	HW		LW	HW	LW	HW		LW	HW
South	CAS	0.13	0.04	−68%	6	9	0.11	0.03	−71%	6	8
	ESCS	0.14	0.06	−59%	6	19	0.13	0.05	−57%	6	18
North	CAS	0.11	0.05	−58%	12	22	0.09	0.03	−63%	12	22
	ESCS	0.10	0.05	−52%	12	22	0.08	0.05	−36%	12	22
East	CAS	0.08	0.04	−50%	5	12	0.07	0.03	−54%	4	12
	ESCS	0.08	0.04	−58%	4	12	0.08	0.03	−65%	4	12
West	CAS	0.20	0.06	−68%	4	18	0.18	0.06	−68%	5	18
	ESCS	0.19	0.05	−74%	4	12	0.18	0.05	−69%	5	12
Building	CAS	0.14	0.06	−58%	12	12	0.11	0.04	−62%	12	12
	ESCS	0.14	0.08	−45%	12	21	0.10	0.06	−46%	12	21

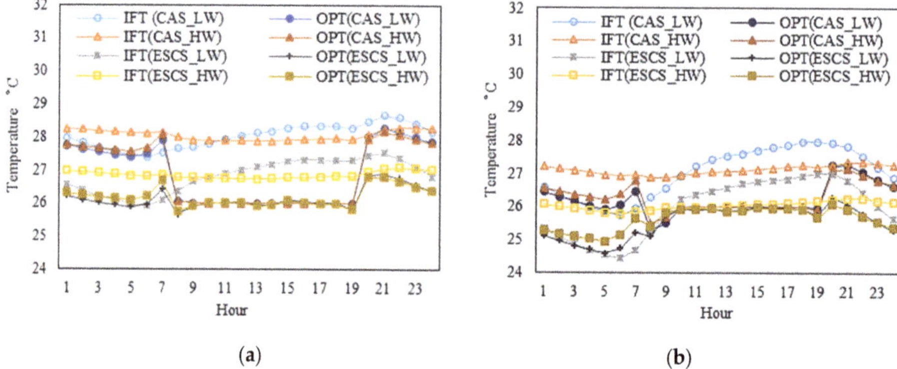

Figure 8. Indoor temperature in the north perimeter zone. (**a**) Indoor temperature on a Beijing cooling design day. (**b**) Indoor temperature on a typical Nanjing day in transient season. (IFT—indoor face temperature, OPT—operative temperature).

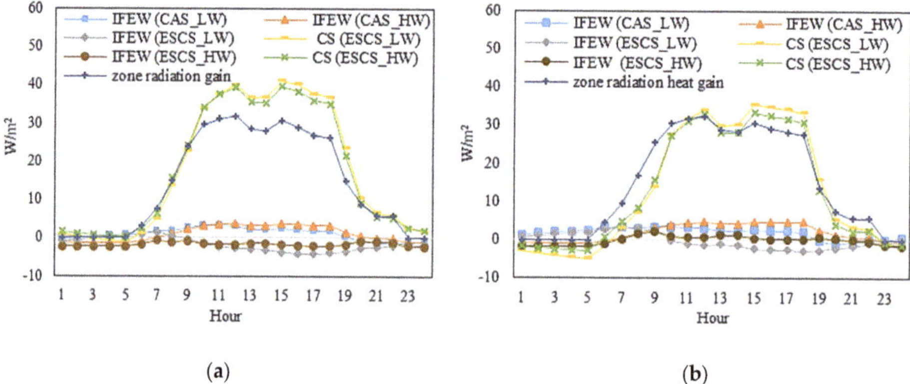

Figure 9. Radiation heat gain in north perimeter zones. (**a**) Radiation heat gain on Beijing cooling design day, (**b**) Radiation heat gain on a typical Nanjing day in transient season. (IFEW—inside face of external wall, CS—cooling ceiling.)

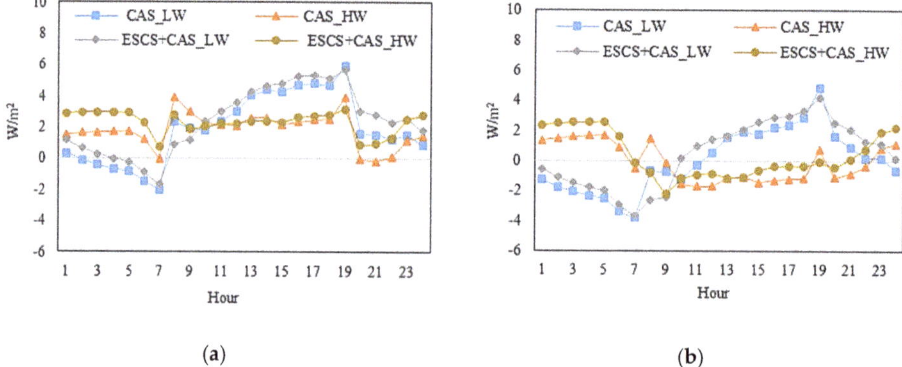

Figure 10. Conduction heat gains on the inside faces of external walls in north perimeter zones. (**a**) Conduction heat gain on a Beijing cooling design day. (**b**) Conduction heat gain on a typical Nanjing day in transient season.

In contrast to the performances during the occupied period, more transmission loads are observed on the external walls with heavy weight structures during the unoccupied period, compared to the loads on the walls of a light weight in the zones with same cooling systems. It is due to the fact that the inside face temperatures of the external surfaces with heavy weight structures still stay higher than the corresponding room operative temperatures in this time; meanwhile, the inside face temperatures of the surfaces with lightweight structures decrease and approach the corresponding operative temperatures after the midnight. For a same structure, a wider gap exists between the inside face temperature and the corresponding operative temperature in the zone with an ESCS in comparison with the performance in that zone with a CAS. More radiation heat transfers occur on the surfaces in the zones with ESCSs, and the consequent conduction heat gains are higher in most of the unoccupied period, and the accumulated values (Figure 10a).

3.1.2. Performance on Typical Days in the Transient Season

On typical days in the transient season, the outdoor dry-bulb temperature ranges surrounding the rooms set the temperatures in Beijing and Nanjing. For a surface with a lightweight structure, the inside face temperature fluctuates with outdoor temperature both in a zone with an ESCS and a zone with an equivalent CAS (Figure 8b). Both the inside face temperatures are higher than the corresponding zones' operative temperatures during occupied period. The maximum conduction gains on these inside faces are close, occurring at 19:00 (Figure 10b). However, the heat fluxes transfer from the inside faces to the outside in most of the unoccupied period, because the inside face temperatures decrease sharply as the outdoor temperatures fall down, and are even lower than the corresponding operative temperatures after midnight. In contrast, for the surfaces with heavy weight structures, the inside face temperatures have little fluctuation. The conduction heat gains on the inside faces are very minor, or even negative during the occupied period, but they increase as time goes on. The maxima occur at 5:00 when the zone operative temperatures are out of control (Figure 10b). The phase lag times are prolonged by 10 hours compared to the performances on the surfaces with lightweight structures. Thus, the 24-hour conduction heat transfers on the external surfaces can be also regarded as the process of cooling charging when the cooling systems operate, and discharging in the rest time. That is the main reason why the values of relative effect of the thermal mass (R) are enhanced on typical days in the transient season.

Although almost all radiation heat gain in the zone with an ESCS can be extracted by a cooling surface through direct or indirect radiation heat transfer during the occupied period, the portion of conduction heat gain on the inside face of the external surface balanced by radiation is not as much as the one on a cooling design day (Figure 9b), average 30% on a surface with lightweight structure and less than 10% on a heavy wall. This is due to the fact that the inside face temperatures of the external surfaces decrease as the outdoor weather becomes cooler. Even so, the inside face temperatures in the zones with ESCSs are relatively lower than the values on the surfaces with the same structures in the zones with CASs (Figure 8b). The difference in heat transfer process between the zones with the different cooling systems leads to an interesting phenomenon on typical days in the transient season: the operative temperatures increase significantly when the CASs are switched off, and stay above the 26 °C during un-occupied period, whereas in the zones with ESCSs the operative temperatures rise slightly and then fall down after the internal heat gains completely disappear. Thus, similarly to the statement in [5,50], the application of night ventilation to cool down a surface with an interior massive layer could be feasible in a zone with a CAS. It may result in a considerable reduction in conduction gains, but it is not necessarily for a zone with a cooling surface. In addition, since a bigger difference exists between the inside face temperature and zone operative temperature in the zone with an ESCS, more conduction heat gain is observed by comparing it with the gain in the zone with an equivalent CAS. The values of relative effect of thermal mass (R) in the perimeter zones with ESCSs become less than the ones in the zones with CASs on the typical days in the transient season.

3.2. The Effect of Thermal Mass Position on System Performance

Preceding research [13–17,20] indicated that the insulation layer should be placed inside when the cooling system runs intermittently. Thus, the external wall structure can be rearranged (i.e., outside concrete + inside insulation). The corresponding thermal mass performances are discussed based on identical thermal environments in the same office building equipped with the combined system. The peak sensible cooling load decreases by approximately 2% to 3% as the thickness of the concrete layer extends from 0 to 200 mm, whereas the corresponding accumulated cooling loads for 24 h change minimally as the thermal mass increases.

Taking the performances in the north perimeter zones with ESCSs as examples, the inside face temperature on the surface with inside insulation approximates to the temperature on the surface with a massive layer inside during the occupied period on the Beijing cooling design day (Figure 11). The conduction gains at these inside faces are also close (Figure 12). Since the conduction heat gain only accounts for a small portion of the total gain in the thermal zone, the peak room cooling load reduces slightly as the thickness of outside concrete layer increases on the cooling design day. However, the accumulated conduction heat gain for the wall with inside insulation is higher than that for the surface with a massive layer inside, because the inside face temperature rises more significantly as the cooling system is turned off.

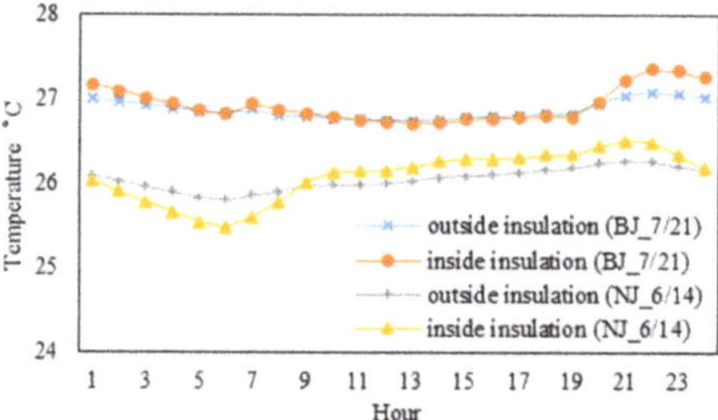

Figure 11. Inside face temperatures of external opaque walls with different structures in north perimeter zones with ESCS.

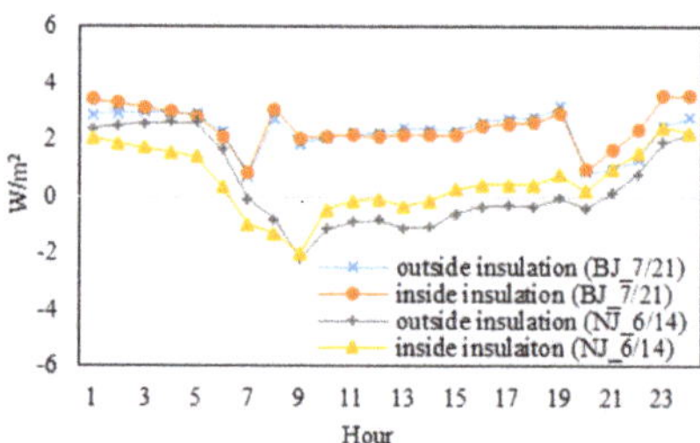

Figure 12. Conduction heat gains at the inside faces of external opaque walls in north perimeter zones with ESCS.

On a typical Nanjing day in June, more conduction gain exists on the external wall with inside insulation during the occupied period compared to the gain of the wall with outside insulation (Figure 12), because little internal heat gain is absorbed by the inside insulation layer, and the portion distributed to the external wall immediately becomes the cooling load through convection and radiation heat transfers. In a typical office building, internal heat gain from occupants, lighting, and electrical equipment accounts for 60% or more of the total gain. Correspondingly, the inside face temperature on a wall with inside insulation is higher than that of the wall with a massive layer inside, most times, on the typical day in June. Although the temperature falls down after midnight, and the corresponding conduction gain decreases and becomes less than the gain on the surface with a massive inside layer, it cannot compensate the excess gain from 09:00 to 23:00 during the day. Therefore, the structure (outside massive layer + inside insulation) has little positive impact on the saving of heat transmission. The accumulated conduction gain on the inside face is close or even higher than the amount on the surface without concrete layer when the degree-hours are absolutely the same.

4. Discussion

Generally, a cooling source with a relatively higher temperature can be utilized by a radiant system, and the entire system coefficient of performance (COP) is expected to be better than a conventional CAS. The present authors investigated a practical TABS operation, and found that the supply water temperature can increase to some degree to maintain the indoor environment at an acceptable comfort level by utilizing the thermal mass and prolonging the radiant system operation time [35]. In addition, from the performances on the typical days in the transient season, the transmission loads caused by an outdoor environment are not as much as the loads on the cooling design days, and the heat transfer in the inside massive layer can be viewed as a process of cooling charging and discharging. Therefore, a potential system strategy, OPPN, could be tried in the zone with ESCS. In the strategy, the auxiliary CAS still runs continuously during the occupied period; the hydronic system starts at midnight and runs continuously with a constant flow rate until the end of the occupied period. The supply water temperature is set at a constant value according to the average outdoor temperature over 24 h and the hydronic system operation hours. Thus, the room operative temperature cannot stay constant but should be in an acceptable range.

As illustrated by Figure 13, during the period from midnight to the earliest occupied time, the inside face temperature of the external wall is relatively higher than the operative temperature in the zone (with ESCS in OPPN) on a Beijing cooling design day. As a consequence, the conduction gain on the inside face is considerably higher than the performance in strategy OPCT (Figure 14), and it is the process for cooling conservation. During the occupied period, the inside face temperature approaches to the zone operative temperature, and the conduction is as low as zero, meaning that the conserved cooling releases to compensate the heat gain on this surface. Thus, the maximum heat gain on the cooling surface in strategy OPPN is not as much as the peak value in OPCT. Considering the effect of additional thermal mass on other structures in the zone, the supply water temperature can be raised from 16 °C (in OPCT) to 20 °C (in OPPN), and the peak cooling load of radiant system can fall down 28% in turn. In addition, the strategy OPPN can be also tried in the zone where the external wall has no concrete layer. However, as the Figure 14 illustrates, little cooling can be conserved in the external wall, and the instantaneous transmission load is still close to the load in the same zone with ESCS in OPCT during occupied period. However, the risk of condensation should be avoided on the cooling surface by some measures when the single hydronic system runs in night.

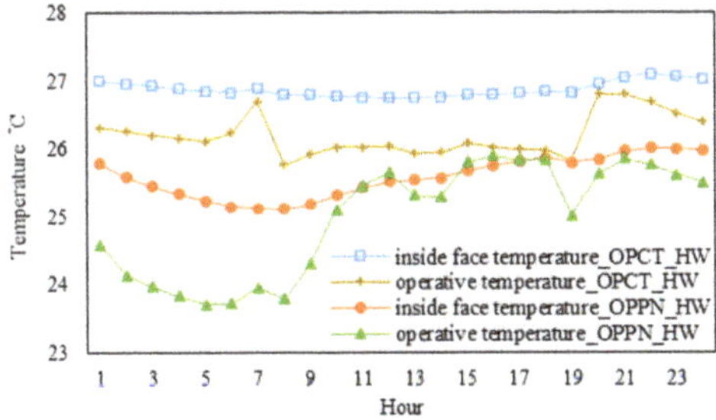

Figure 13. Indoor temperatures in the zones with ESCS in different operation strategies on a Beijing cooling design day.

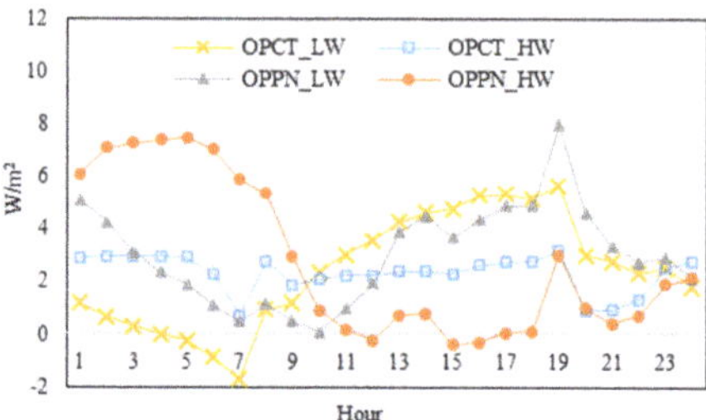

Figure 14. Conduction heat gains on the inside faces of external walls in north perimeter zones with ESCS on Beijing cooling design day.

Therefore, the cooling water temperature of the radiant system could be raised by taking the effect of internal thermal mass and improving the system's operation strategy. Some low-grade cooling energy sources can be directly used for free cooling, such as geothermal systems. The geothermal system has been applied in several projects where radiant heating and cooling is employed [32,36,51].

5. Conclusions

The effects of thermal mass in external walls were investigated by simulating the energy performances in a typical office building, rather than considering only the heat transfer for an individual structure. Operative temperature is employed to evaluate the thermal comfort level in both zones—those having a combined system and an equivalent CAS alone. The simulation tool of Energy-Plus is employed in the research, and it is based on heat balance method. It takes the influences of human actives and the cooling system operation strategy into account, in addition to the impacts of building physics and climatic conditions. Thus, the study results approximate real life and muse be the references for building and radiant cooling system designs, although the computation process is relatively complex.

The article introduces a new concept, relative effect (R), and takes the performance in the northern perimeter zone as an example with which to quantificationally distinguish the heat transfer process in the zone with an ESCS from the performance in the zone with an equivalent CAS, confirming that the cooling surface can decrease the heat storage capacity of the building envelope by radiation heat transfer [29]. A big portion of conduction gain on an inside face of external walls is balanced by radiation heat transfer during the occupied period on cooling design days (e.g., 66% and 88%, respectively for the zone with heavy weight and the zone with light weight in the north zone). The research shows some new findings as follows.

- The peak cooling load and the accumulated load of the combined system are higher than the corresponding values of the equivalent CAS by 9%–11% and 3%–4%, respectively, in the buildings of the same structure. The results are basically consistent with the findings from laboratory experiments [49].
- Compared to the performance in the zones with equivalent CASs, the effect of thermal mass is more evident in the zones with ESCSs on the cooling design days. The values of relative effect (R) are 22%–31% and 34%–40%, respectively, for the building with CASs and the building with ESCSs. The relative effects are about 60% and 45%, respectively, on the typical days in transient season.
- An external wall with a massive inside layer is suggested for a zone with a radiant system either in intermittent operation or in continuous operation, because the inside insulation may lead to a greater transmission load, especially in a case with high internal heat gain (>60% of the total gain). The inside massive layer can also be used for cooling conservation in a different operation strategy (OPPN), and the peak cooling load of ESCS decreases 28%.

Based the results mentioned above, maintaining the thermal mass with a certain weight can be a key measure for a low-carbon building or green building, particularly in the zones equipped with radiant systems. In addition, the risk of condensation on cooling surfaces should be avoided, especially when the radiant system works in the situation without a dehumidification system running. In future work, some tests are going to be carried out to determine instantaneous heat fluxes on a cooling surface, and verify the interaction between a cooling surface and its surroundings.

Author Contributions: Conceptualization, J.N.; methodology, R.H.; software, R.H.; writing—original draft preparation, R.H.; wiring—review and editing, G.L.; funding acquisition, R.H. All authors have read and agreed to the published version of the manuscript.

Funding: This research was funded by Guangxi Natural Science Foundation, grant number 2018GXNSFBA050022; and funded by Guangxi Basic Research Ability Improvement Foundation, grant number 2019KY0220.

Acknowledgments: The project is supported by Guangxi Natural Science Foundation (number 2018GXNSFBA050022), and Guangxi basic research ability improvement Foundation (number 2019KY0220).

Conflicts of Interest: The authors declare there is no conflicts of interest regarding the publication of this paper.

Nomenclature and Definition:

XPS	Extrusion polystyrene insulation
q_{conv}	Convective h eat transfer from surfaces in a room W
q''_{conv}	Convective heat flux to zone air from a surface W
q_{IV}	Sensible load caused by infiltration and ventilation W
q''_{SW}	Net short wave radiant flux to surface from lights W
q''_{LWS}	Long wave radiant flux from equipment in zone W
q''_{LWX}	Long wave radiant flux exchange between surfaces W
q''_{sol}	Transmitted solar radiant flux absorbed at surface W
q''_{ki}	Conductive flux through the inside face of surface W
q''_{k-ina}	Conductive flux through inside face of inactive surface W
$\sum_{24h} q''_{ki}$	Accumulated conductive flux through the inside face of a surface for 24 hours on a typical day W·h
q_{CE}	Convective parts of internal loads W
q_{cs}	Heat extracted by active surfaces W
q_c	Conduction heat transfer of inside face of ceiling W
q_f	Conduction heat transfer of inside face of floor W
q_{ZH}	Heat extraction from a thermal zone W
$q_{air\ sys}$	Sensible cooling load of air-conditioning system W
q_{ZRH}	Instantaneous zone radiation heat gain W
q_{CSRH}	Radiation heat gain at the active surfaces W
q_{ZT}	Room sensible cooling load handled by combined system W
q''_{ko}	Conductive flux through the outside face of surface W
$\sum_{24h} q_{ZT}$	Accumulated room sensible cooling load by combined system for 24 hours on a typical day W·h

References

1. Feng, Y. Thermal design standards for energy efficiency of residential buildings in hot summer/cold winter zones. *Energy Build.* **2004**, *36*, 1309–1312. [CrossRef]
2. Kaynakli, O. A review of the economical and optimum thermal insulation thickness for building applications. *Renew. Sustain. Energy Rev.* **2012**, *16*, 415–425. [CrossRef]
3. China Academy of Building Sciences. *Design Standard for Energy Efficiency of Public Buildings(GB50189-2015)*; China Construction Industry Press: Beijing, China, 2015. (In Chinese)
4. Balaras, C.A. The role of thermal mass on the cooling load of buildings: An overview of computational methods. *Energy Build.* **1996**, *24*, 1–10. [CrossRef]
5. Shaviv, E.; Yezioro, A.; Capeluto, I.G. Thermal mass and night ventilation as passive cooling design strategy. *Renew. Energy* **2001**, *24*, 445–452. [CrossRef]

6. Lee, K.H.; Joo, M.C.; Baek, N.C. Experimental evaluation of simple thermal storage control strategies in low-energy solar houses to reduce electricity consumption during grid on-peak periods. *Energies* **2015**, *8*, 9344–9364. [CrossRef]

7. Albayyaa, H.; Hagare, D.; Saha, S. Energy conservation in residential building by incorporating passive solar and energy efficiency design strategies and higher thermal mass. *Energy Build.* **2019**, *182*, 205–213. [CrossRef]

8. Bahrar, M.; Djamai, Z.L.; Mankibi, M.E.; Larbi, A.S.; Salvia, M. Numerical and experimental study on the use of microencapsulated phase change materials (PCMs) in textile reinforced concrete panels for energy storage. *Sustain. Cities Soc.* **2018**, *41*, 455–468. [CrossRef]

9. Zhu, N.; Li, S.; Hu, P.; Wei, S.; Deng, R.; Lei, F. A review on applications of shape-stabilized phase change materials embedded in building enclosure in recent ten years. *Sustain. Cities Soc.* **2018**, *43*, 251–264. [CrossRef]

10. Ferrari, S.; Beccali, M. Energy-environmental and cost assessment of a set of strategies for retrofitting a public building toward nearly zero-energy building target. *Sustain. Cities Soc.* **2017**, *32*, 226–234. [CrossRef]

11. Stritih, U.; Tyagi, V.V.; Stropnik, R.; Paksoy, H.; Haghighat, F.; Joybari, M. Integration of passive PCM technologies for net-zero energy buildings. *Sustain. Cities Soc.* **2018**, *41*, 286–295. [CrossRef]

12. Al-Sanea, S.A. Evaluation of heat transfer characteristics of building wall elements. *J. King Saud Univ. Eng. Sci.* **2000**, *12*, 285–313. [CrossRef]

13. Al-Sanea, S.A.; Zedan, M.F. Effect of insulation location on initial transient thermal response of building walls. *J. Therm. Envel. Build. Sci.* **2001**, *24*, 275–300. [CrossRef]

14. Al-Sanea, S.A.; Zedan, M.F.; Al-Hussain, S.N. Effect of thermal mass on performance of insulated building walls and the concept of energy saving potential. *Appl. Energy* **2012**, *89*, 430–442. [CrossRef]

15. Tsilingiris, P.T. The influence of heat capacity and its spatial distribution on the transient wall thermal behavior under the effect of harmonically time-varying driving forces. *Build. Environ.* **2006**, *41*, 590–601. [CrossRef]

16. Tsilingiris, P.T. Wall heat loss from intermittently conditioned spaces-The dynamic influence of structural and operational parameters. *Energy Build.* **2006**, *38*, 1022–1031. [CrossRef]

17. Tsilingiris, P.T. Parametric space distribution effects of wall heat capacity and thermal resistance on the dynamic thermal behavior of walls and structures. *Energy Build.* **2006**, *38*, 1200–1211. [CrossRef]

18. Deng, J.; Yao, R.; Yu, W.; Zhang, Q.; Zhan, B. Effectiveness of the thermal mass of external walls on residential buildings for part-time Part-space heating and cooling using the state method. *Energy Build.* **2019**, *190*, 155–171. [CrossRef]

19. Kumar, S.; Singh, M.K.; Mathur, A.; Mathur, S.; Mathur, J. Thermal performance and comfort potential estimation in low-rise high thermal mass naturally ventilated office buildings in India: An experimental study. *J. Build. Eng.* **2018**, *20*, 569–584. [CrossRef]

20. Reilly, A.; Kinnane, O. The impact of thermal mass on building energy consumption. *Appl. Energy* **2017**, *198*, 108–121. [CrossRef]

21. Rodriguez-Rodriguez, I.; Gonzalez Vidal, A.; Ramallo Gonzalez, A.; Zamora, M. Commissioning of the controlled and automatized testing facility for human behavior and control (CASITA). *Sensors* **2018**, *18*, 2829. [CrossRef]

22. Eben Saleh, M.A. Thermal insulation of buildings in a newly built environment of a hot dry climate: The Saudi Arabian experience. *Int. J. Ambient Energy* **1990**, *11*, 157–168. [CrossRef]

23. Eben Saleh, M.A. Impact of thermal insulation location on buildings in hot dry climates. *Sol. Wind Technol.* **1990**, *7*, 393–406. [CrossRef]

24. Kossecka, E.; Kosny, J. The effect of insulation and mass distribution in exterior walls on the dynamic thermal performance of whole buildings. In Proceedings of the Thermal Performance of the Exterior Envelopes of Buildings VII, Clearwater Beach, FL, USA, 6–10 December 1998; pp. 721–731.

25. Kossecka, E.; Kosny, J. Influence of insulation configuration on heating and cooling loads in a continuously used building. *Energy Build.* **2002**, *34*, 321–331. [CrossRef]

26. Verbeke, S.; Audenaert, A. Thermal inertia in buildings: A review of impacts across climate and building use. *Renew. Sustain. Energy Rev.* **2018**, *82*, 2300–2318. [CrossRef]

27. Liu, X.H.; Jiang, Y. *Temperature and Humidity Independent Control Air-Conditioning System*, 1st ed.; China Building Industry Press: Beijing, China, 2006. (In Chinese)

28. Olsthoorn, D.; Haghighat, F.; Moreau, A.; Lacroix, G. Abilities and limitations of thermal mass activation for thermal comfort, peak shifting and shaving. *A Rev. Build. Environ.* **2017**, *118*, 113–127. [CrossRef]

29. Niu, J.L.; Kooi, J.; Ree, H. Energy saving possibilities with cooled-ceiling systems. *Energy Build.* **1995**, *23*, 147–158. [CrossRef]

30. Feng, J.D.; Schiavon, S.; Bauman, F. Cooling load differences between radiant and air systems. *Energy Build.* **2013**, *65*, 310–321. [CrossRef]

31. Feng, J.D.; Bauman, F.; Schiavon, S. Experimental comparison of zone cooling load between radiant and air systems. *Energy Build.* **2014**, *84*, 152–159. [CrossRef]

32. Hu, R.; Niu, J.L. A review of the application of radiant cooling & heating systems in Mainland China. *Energy Build.* **2012**, *52*, 11–19.

33. Romanía, J.; Graciab, A.; Cabezaa, L.F. Simulation and control of thermally activated building systems (TABS). *Energy Build.* **2016**, *127*, 22–42. [CrossRef]

34. Haniff, M.F.; Selamat, H.; Yusof, R.; Buyamin, S. Fatimah Sham Ismail, Review of HVAC scheduling techniques for buildings towards energy-efficient and cost-effective operations. *Renew. Sustain. Energy Rev.* **2013**, *27*, 94–103. [CrossRef]

35. Hu, R.; Niu, J.L. Operation Dynamics of Building with Radiant cooling System Based on Beijing Weather. *Energy Build.* **2017**, *151*, 344–357. [CrossRef]

36. Tahersima, M.; Tikalsky, P.; Revankar, R. An experimental study on using a mass radiant floor with geothermal system as thermal battery of the building. *Build. Environ.* **2018**, *133*, 8–18. [CrossRef]

37. Laouadi, A. Development of a radiant heating and cooling model for building energy simulation software. *Build. Environ.* **2004**, *39*, 421–431. [CrossRef]

38. ASHRAE. Thermal comfort (Chapter 9). In *Fundamental*, SI ed.; American Society of Heating, Air-Conditioning and Refrigeration Engineers: Atlanta, GA, USA, 2017.

39. ASHRAE. Radiant heating and cooling (Chapter 55). In *Handbook, Heating, Ventilation and Air-Conditioning Application*; SI Version; American Society of Heating, Air-Conditioning and Refrigeration Engineers: Atlanta, GA, USA, 2019.

40. Causone, F.; Corgnati, S.P.; Filippi, M.; Olesen, B.W. Solar radiation and cooling load calculation for radiant systems: Definition and evaluation of the Direct Solar Load. *Energy Build.* **2010**, *42*, 305–314. [CrossRef]

41. ASHRAE. *Handbook Fundamentals, SI ed.*; American Society of Heating, Refrigerating and Air-Conditioning Engineering, Inc.: Atlanta, GA, USA, 2009.

42. Rhee, K.N.; Kim, K.W. A 50 year review of basic and applied research in radiant heating and cooling systems for the built environment. *Build. Environ.* **2015**, *91*, 166–190. [CrossRef]

43. Energy-Plus. Available online: https://www.energyplus.net/sites/all/modules/custom/nrel_custom/pdfs/pdfs_v9.2.0/EngineeringReference.pdf (accessed on 1 October 2019).

44. China Architecture Standard Design and Research Institute; China Academy of Building Sciences. *Code for Design of Civil Buildings (GB 50352-2019)*; China Construction Industry Press: Beijing, China, 2019. (In Chinese)

45. Jingran, M.; Qin, S.J.; Li, B.; Salsbury, T. Economic model predictive control for building energy systems. In Proceedings of the sofISGT2011: Innovative smart grid technologies, Anaheim, CA, USA, 17–19 January 2011; pp. 1–6.

46. Lee, K.H.; Braun, J.E. Development of methods for determining demand-limiting set-point trajectories in buildings using short-term measurements. *Build. Environ.* **2008**, *43*, 1755–1768. [CrossRef]

47. Lee, K.H.; Braun, J.E. A data driven method for determining zone temperature trajectories that minimize peak electrical demand. *ASHRAE Trans.* **2008**, *114*, 65–74.

48. Lee, K.H.; Braun, J.E. Evaluation of methods for determining demand-limiting set-point trajectories in buildings using short-term measurements. *Build. Environ.* **2008**, *43*, 1769–1783. [CrossRef]

49. Woolley, J.; Schiavon, S.; Bauman, F.; Raftery, P.; Pantelic, J. Side-by-side laboratory comparison of space heat extraction rates and thermal energy use for radiant and all-air systems. *Energy Build.* **2018**, *176*, 139–150. [CrossRef]

50. Guo, R.; Hu, Y.; Liu, M.P. Heiselberg, Influence of design parameters on the night ventilation performance in office buildings based on sensitivity analysis. *Sustain. Cities Soc.* **2019**, *50*, 101661. [CrossRef]

51. Zhang, L.H.; Huang, X.K.; Liang, L.; Liu, J.Y. Experimental study on heating characteristics and control strategies of ground source heat pump and radiant floor heating system in an office building. *Procedia Eng.* **2017**, *205*, 4060–4066. [CrossRef]

© 2020 by the authors. Licensee MDPI, Basel, Switzerland. This article is an open access article distributed under the terms and conditions of the Creative Commons Attribution (CC BY) license (http://creativecommons.org/licenses/by/4.0/).

Article

Modeling and Forecasting End-Use Energy Consumption for Residential Buildings in Kuwait Using a Bottom-Up Approach

Turki Alajmi [1] and Patrick Phelan [2,*]

[1] Energy and Building Research Center, Kuwait Institute for Scientific Research, Kuwait City 13109, Kuwait; tajmi@kisr.edu.kw

[2] School for Engineering of Matter, Transport and Energy, Arizona State University, Tempe, AZ 85287-6106, USA

* Correspondence: phelan@asu.edu; Tel.: +1-480-965-1625

Received: 29 February 2020; Accepted: 14 April 2020; Published: 17 April 2020

Abstract: To meet the rapid-growing demand for electricity in Kuwait, utility planners need to be informed on the energy consumption to implement energy efficiency measures to manage sustainable load growth and avoid the high costs of increasing generation capacities. The first step of forecasting the future energy profile is to establish a baseline for Kuwait (i.e., a business-as-usual reference scenario where no energy efficiency incentives were given and the adoption of energy efficient equipment is purely market-driven). This paper presents an investigation of creating a baseline end-use energy profile until 2040 for the residential sector in Kuwait by using a bottom-up approach. The forecast consists of mainly two steps: (1) Forecasting the quantity of the residential energy-consuming equipment in the entire sector until 2040 where this paper used a stock-and-flow model that accounted for the income level, electrification, and urbanization rate to predict the quantify of the equipment over the years until 2040, and (2) calculate the unit energy consumption (*UEC*) for all equipment types using a variety of methods including EnergyPlus simulation models for cooling equipment. By combining the unit energy consumption and quantity of the equipment over the years, this paper established a baseline energy use profile for different end-use equipment for Kuwait until 2040. The results showed that the air conditioning loads accounted for 67% of residential electrical consumption and 72% of residential peak demand in Kuwait. The highest energy consuming appliances were refrigerators and freezers. Additionally, the air conditioning loads are expected to rise in the future, with an average annual growth rate of 2.9%, whereas the lighting and water heating loads are expected to rise at a much lower rate.

Keywords: energy modeling; bottom-up models; building archetype simulation; unit energy consumption; end-use forecasting; diffusion rate

---

1. Introduction

Kuwait has experienced a steady increase in its population since the 1960s, however, with the turn of the century, an exponential rise has been observed as per Figure 1 [1]. This steep increase, along with economic growth, has resulted in higher electrical consumption, exceeding approximately 30 TWh per annum since 2000, whereas the highest level in the 1980s was less than 10 TWh [2]. Aside from the high population growth and rise in new construction, Kuwait also has a high energy use per capita, as shown in Figure 2 [3], which is mainly driven by the heavy subsidization of the cost of electricity. Having more than doubled since the early 1990s, per capita energy consumption poses a serious problem [2]. Considering the demand for labor and the fast-paced development trend in the region, both Figures 1 and 2 clearly indicate the impact created on the electrical load for Kuwait. In addition, according to the Ministry of Energy and Water, the peak demand is expected to reach 30,000 MW by 2030, whilst 70% of this is attributed to new residential construction [4].

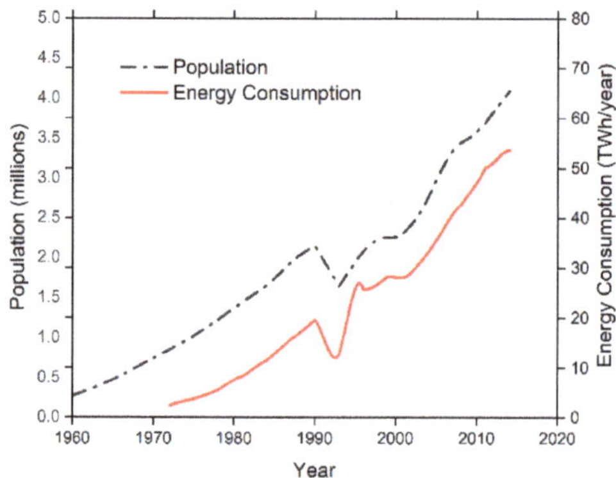

Figure 1. Population and electricity growth trends in Kuwait from 1960 to 2015 [1].

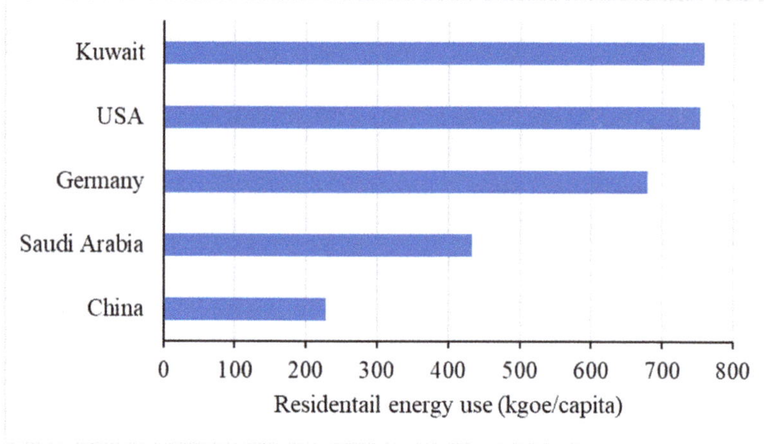

Figure 2. Residential energy use per capita in 2014 (kg of oil equivalent per capita) [3].

Given the growing population and new construction initiatives in the form of housing subsidies coupled with high energy consumption per capita, energy consumption growth trends create a risk for the stability of the electrical grid and meeting the national demand. While extensive studies have been published on building energy use in Kuwait, most have been observed to be geared toward the evaluation of certain policies or retrofit programs related to energy efficiency. In the literature, end-use energy consumption for residential buildings in Kuwait has been identified in studies that utilize archetypes. Baqer and Krarti [5] modeled a prototypical Kuwaiti villa and carried out a series of analyses to ascertain the effectiveness of certain energy policies, and the impact of various energy efficiency measures on energy use and peak demand. It was observed that air conditioning accounts for 72% of the total electrical usage, whereas lighting and miscellaneous household appliances account for 22% of the energy consumption combined.

Another study conducted by Krarti and Hajiah [6] examined the impact of daylight time savings (DST) on energy use for various types of buildings. Similarly, the analysis was based on a series of archetypical models that represented buildings in the residential and commercial sectors. According to their results, space cooling represents a majority of the usage and peak demand at 48% of annual energy use, and represents a peak load of 64%.

To forecast energy demand, a study by Wood and Alsayegh [7] modeled the electrical demand up to 2030 by using a top-down approach. It was developed based on historic data of oil income, gross domestic product (GDP), population, and electric load. However, a forecasting model of the energy consumption and demand by end-use using a bottom-up approach has not, to the best of our knowledge, been developed as yet. Should a breakdown of energy end-uses be analyzed and forecasted, better building energy use can be strategized as well as the development of more effective codes and standards. Given that 57% of the energy consumption is attributed to the residential sector, it is crucial to assess the baseline energy consumption patterns [6].

A number of different algorithms are available to study the residential energy consumption [8–12]. These models depend on accurate input data to generate meaningful results. Generally, the analysis methods can be divided into "top-down" and "bottom-up" approaches, as shown in Figure 3 [13,14]. The top down approach calculates the energy consumption for the entire target sector by using the econometric and technological data for the region [15,16]. On the other hand, the bottom up approach calculates the individual building energy consumption by using either statistical model or engineering models that are then aggregated to obtain the energy use for the entire sector [17,18].

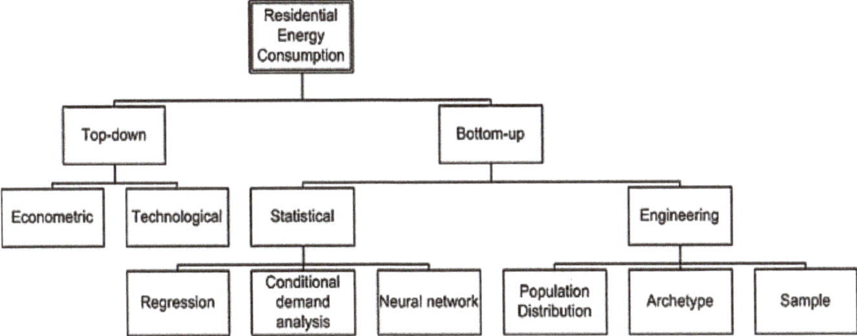

Figure 3. Modeling techniques to estimate the residential energy consumption. Reprint with permission [14]; Copyright 2009, Elsevier.

Statistical and engineering methods represent two distinct approaches applied in the bottom-up models to determine the energy consumption of specified end-uses [14]. The statistic method first identifies a sample of households that represent the entire building stock and then uses regression and other statistic models to predict the energy use of the sampled household and hence the entire building stock [19]. Energy modeling is gaining more popularity in the bottom-up approach with the development of energy simulations. This approach utilizes the archetype models to represent the building stock and aggregate the calibrated model results to predict the energy consumptions of the entire building stock [14]. One major advantage of the energy model is that it can predict the end-use distribution without requiring sub-metering. This offers great flexibility and more detail in terms of the end-use characteristics when compared to the statistical model. However, to obtain accurate simulation results, a high-quality set of inputs often from onsite surveys and calibration to the historical energy use data are required.

2. Residential Characteristics in Kuwait

Buildings account for nearly one-third of the global final energy consumption and 55% of global electricity demand. Electricity demand growth in buildings has been particularly rapid over the last 25 years, accounting for nearly 60% of total growth in global electricity consumption [20]. Similar to the global figures, the energy demand in Kuwait is also rising rapidly. Due to the hot climate and high energy use per capita, residential buildings account for a significant portion of the total electrical consumption. Mainly attributed to the air conditioning and refrigeration loads, 57% of the peak demand consumed in Kuwait is from the residential sector [21].

It should also be underlined that the main factor contributing to such high rates of energy-use-per-capita is subsidization. The government in Kuwait subsidizes 94.7% of the total cost of electricity, leaving a factional cost of only 2 fils/kWh ($0.007/kWh) for the end-user [5]. Targeting electrical consumption in buildings will drastically reduce the impact on the electric grid, but also reduce CO_2 emissions, since Kuwait heavily relies on fossil fuels for generation. In doing so, analyzing the energy consumption patterns is crucial.

Due to the vast differences between Kuwaiti and non-Kuwaiti (i.e., expatriate) residential households, observing both sides separately is very important. The differences range from electrical consumption patterns, appliance ownership rates, and occupancy behavior to different utility rate structures. In further detail, approximately 90% of Kuwaiti households would fall under the single-detached home category, as most are single-family homes, whereas roughly 70% of non-Kuwaiti families would be under the multi-family home category, mostly living in apartment buildings [22]. The electrical rate for the residential sector, which includes single-detached dwellings, is 1.8 fils/kWh ($0.006/kWh). The rate for the investment sector, which includes multi-family homes, is approximately 250% more when compared to the residential sector. Furthermore, the average household size for Kuwaiti homes is roughly eight people, whereas non-Kuwaiti homes are smaller, at around four persons [22,23]. In addition, the average growth rate of the expatriate population between 2000 and 2016 was 5.2%, compared to 2.9% for the Kuwaiti population [1,22].

3. Methodology

This work presents a bottom-up approach for modeling and forecasting end-use energy consumption and demand in Kuwait's residential buildings up until 2040. The methodology relies on information pertaining to the energy consumption of specific household equipment and appliances, where factors such as quantity, operating hours, and power requirements are accumulated and extrapolated to a national scale to ultimately estimate the usage patterns in Kuwait. Therefore, energy consumption and demand are calculated at the individual level and aggregated to estimate the national consumption and demand. In this model, end-uses were broken into air conditioning, lighting, appliances, and space heating and water heating, and further sub-categorized by different technologies. Moreover, each end-use category was further broken down by different equipment and appliances with corresponding data on diffusion rates and energy efficiency ratings. The rate of diffusion was based on data obtained from surveys and the available literature [23,24]. The driver variables of this model were based on macroeconomic variables such as population, household size and income and engineering variables like unit energy consumption, and efficiency ratings. Figures 4 and 5 illustrate the modeling structure.

The initial step is to model the quantity of equipment owned and the present initial stock. The sales and stock turnover are then derived from first purchases and replacements. The first purchases are driven by a growth in population and increase in ownership, while replacements are calculated based on the age of equipment and a retirement function. Next, the average unit energy consumption (*UEC*) and unit power demand (*UPD*) per equipment are derived and the total energy consumption and peak demand are modeled using the following general equations:

$$Total\ Energy\ Consumption(y) = \sum_{i=1}^{L} Stock(y,i) \times UEC(y-i), \tag{1}$$

$$Peak\ Load\ Demand(y) = \sum_{i=1}^{L} Stock(y,i) \times UPD(y-i), \tag{2}$$

where *Stock* (y, i) represents the quantity of equipment of vintage (i) remaining annually in year (y). The variable *UEC* (y, i) on the other hand, denotes the unit energy consumption at the corresponding year of purchase $(y - i)$ and the *UPD* (y, i) is the unit demand power during the peak time. Finally, the overall useful life of the equipment is represented by *L*. Due to the lack of published information, acquiring data on the sales volumes of equipment, efficiency ratings, ownership details, and daily consumption patterns is not at all feasible for the state of Kuwait. This analysis therefore utilized an array of surveys that included national statistics and numerous reports published by the government [2,22,23].

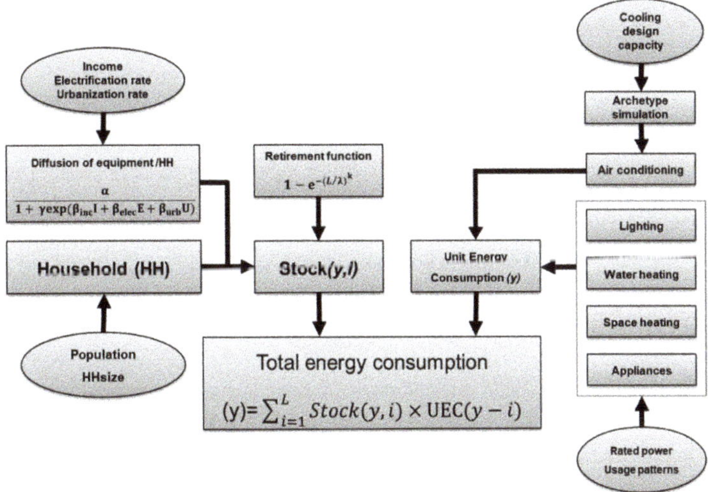

Figure 4. End-use energy consumption model structure.

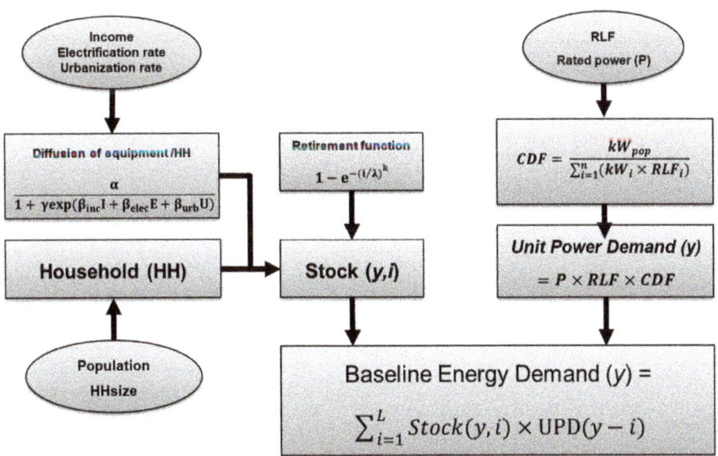

Figure 5. Energy demand model structure.

3.1. Stock and Diffusion Rate

Since the overall consumption of electricity is impacted by the total quantity of equipment, it is crucial to calculate the adoption rates for the population as well as the total sales numbers of end-use equipment. The sales are the sum of initial purchases of equipment and the replacement purchases, which includes replacements-on-burnout and early retirements. The calculations for replacements involve the age of the equipment within the stock and a retirement function, which represents the percentage of failed equipment in a vintage stock:

$$Stock(y) = Sales(y) + Stock(y-1),$$ (3)

$$Sales(y) = First\ purchases(y) + Replacements(y),$$ (4)

First purchases, shown in Equation (4), represent an increase in the stock quantity that can be due to new construction projects such as housing subsidies by the Public Authority of Housing Welfare (PAHW) or an increased rate of equipment diffusion per household, as shown in Equation (5):

$$First\ purchases(y) = H(y)D(y),$$ (5)

where $H(y)$ represents the number of new households based on [22]. $D(y)$ is the equipment diffusion rate per household. Equipment diffusion rates are not available as input data, but are projected according to a macroeconomic model using a logistic function [25,26]:

$$D(y) = \frac{\alpha}{1 + \gamma + e^{-(\beta_1 I(y) + \beta_2 E(y) + \beta_3 U(y))}}$$ (6)

where $I(y)$ denotes the average annual income per household (y), whereas $E(y)$ is the electrification rate, $U(y)$ is the urbanization rate, and γ and β are the parameters for scale. For the case of Kuwait, since the income, electrification, and urbanization rates are relatively high, diffusion rates for equipment are reflective of this phenomenon in the analysis. The logistic function, by definition, has a maximum value of one at which the saturation level is reached. However, some households have more than one appliance or equipment of the same type. Therefore, the logistic function is scaled by the parameter α, as seen in Equation (6), which is the saturation level [25]. As the climate conditions directly impact the air conditioner ownership rates, cooling degree days (CDD) were used instead of an urbanization rate in the equation above to calculate the diffusion rates of AC units. For some appliances, the sale price affects the diffusion rate as purchases depend on affordability. Therefore, a price variable was added for some appliances based on [27]. Replacement stock are attained from previous sales as in Equation (7):

$$Replacements(y) = \sum_{i=1}^{L} Sales(y-i) \times Retirements(i)$$ (7)

In Equation (7), *Retirements* (i) represents the probability of the equipment retiring at a given lifetime for each year up to its entire lifetime (L), and is modeled using a Weibull distribution [16,28]:

$$Retirements(i) = 1 - e^{-(i/\lambda)^k}$$ (8)

where i is the number of years after the equipment is purchased; λ is a scale parameter; and k is a shape parameter, which determines the way the failure rate changes through time. These parameters were estimated for each equipment based on [29].

3.2. Unit Energy Consumption

The next section describes the methods and assumptions for determining the average unit energy consumption (*UEC*) for each piece of equipment. *UEC* depends on the typical product used (size and rated power), the use patterns, and equipment efficiency. Therefore, the *UEC* model includes information on equipment usage and lifetime profiles as well as stock energy efficiencies by vintage and efficiency improvement profiles [5,30]. The assumption of the efficiency improvement of the appliances over time was made based on [31,32], and the likely improvement was 1–5%, depending on the equipment, considering the technical limitation of the technology.

3.2.1. Air Conditioning

Space conditioning is a large driver of energy consumption in residential buildings and is affected by many variables like weather, building envelope efficiency, building size, equipment types, and occupant behaviors. Therefore, it is challenging to determine the *UEC* for AC units and some additional complexity is required for modeling space conditioning in order to obtain reasonable accuracy. This paper used archetype simulation models to estimate the average *UEC* for AC systems in the residential building stock in Kuwait. The simulation models were created in DesignBuilder, which is a user interface for the EnergyPlus simulation engine. The weather dataset used as input for the simulation models was the typical meteorological year (TMY) for Kuwait, as developed by the Kuwait Institute for Scientific Research (KISR) [33]. The TMY datasets represent one year of hourly weather data extracted from long-term data records. The data consisted of the dry-bulb temperature, diffuse radiation, direct normal radiation, wind speed, wind direction, and relative humidity, which were collected from the KISR's weather stations. Four archetype models with different thermal and equipment performance parameters were created to represent the residential building stock in Kuwait. According to the available information, and based on detailed study as part of the Kuwait-MIT (Massachusetts Institute of Technology) projects on the sustainability of Kuwait's built environment [34–36], Table 1 summarizes the archetype parameters used in the simulation. Figure 6 shows the geometry of a sample archetype model. The results from the simulation are shown in Table 2.

Table 1. Archetype parameters [34–36].

Parameters	Archetype			
	A	**B**	**C**	**D**
Construction year	60s–80s	60s–80s	80s–Present	10s–Present
Wall U-value (W/m^2·K)	2.53	2.53	0.62	0.32
Roof U-value (W/m^2·K)	1.56	0.53	0.53	0.40
Window U-value (W/m^2·K)	5.96	2.89	2.89	2.33
Window SHGC	0.86	0.76	0.37	0.65
HVAC COP	2.00	2.20	2.40	2.90
Cooling setpoint temperature (°C)	21	21	21	21
Window-to-wall ratio (%)	20	20	20	20
Infiltration (ACH)	0.80	0.80	0.50	0.30
Occupancy density (Occ/m^2)	0.012	0.012	0.012	0.012

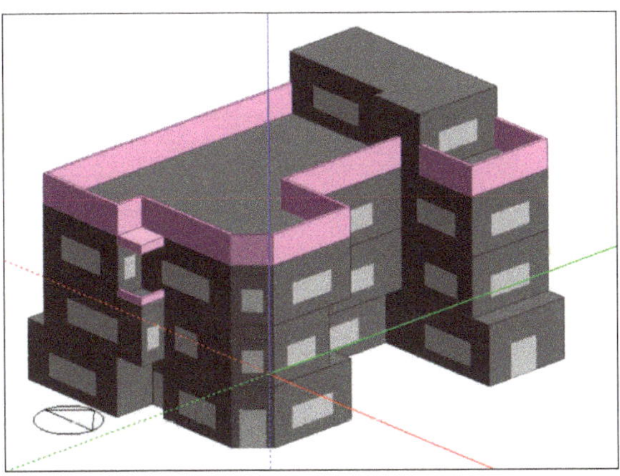

Figure 6. Screenshot of DesignBuilder software interface of a sample archetype model.

Table 2. Average unit energy consumption of AC systems for different residential dwellings in Kuwait.

UEC (kWh/m^2/year)	Archetype			
	A	**B**	**C**	**D**
AC systems	201.08	184.73	130.43	113.57

3.2.2. Water Heaters

The unit energy consumption for a water heater was estimated through Equation (9) [37]:

$$UEC = \frac{Usage \times c_p (T_{supply} - T_{tank})}{EF} \qquad (9)$$

where usage is the household hot water usage in cubic meter per day; c_p is the volumetric specific heat of water (Jm^{-3}K^{-1}); T_{supply} is the incoming cold-water (C); T_{tank} is the tank temperature (C); and EF is the energy factor of the water heater. This was assumed to be 0.904 for standard electric water heaters and 0.95 for high efficiency ones [38]. Electricity is the only fuel used for water heating in residential buildings in Kuwait.

3.2.3. Lighting

Since all electrified households use electricity for lighting, the model assumes that lighting diffusion is equal to the national electrification rate, which is almost 100% for Kuwait [1]. However, the lighting energy is largely determined by the number of lighting fixtures, type of lamps, and usage patterns. Therefore, the residential lighting stock was broken down by lamp type, based on the 2010 lighting stock data in Kuwait [39,40]. Almost 50% of the lighting stock in Kuwait is incandescent bulbs and around 37% is compact fluorescent lamps (CFL). The daily average use is estimated to be seven hours based on [30].

3.2.4. Appliances

The home appliance end-use in residential households includes electric appliances like refrigerators, televisions, computers, and others. The *UEC* for appliances is the product of the nameplate wattage and the usage hours. For products with multiple modes like standby mode, energy consumption for each mode is calculated separately and added to obtain the total energy consumption in all modes. The average hour use, rated wattage, and life span for most of the appliances was estimated based on [23,29,30]. Table 3 lists the various metrics that can be used to calculate the modeled energy usage broken down by appliances.

Table 3. List of household appliance metrics that include corresponding power requirements, average run-time, and useful life.

Appliance	Rated Power (W)	UEC (kWh/year)	Usage (hours/week)	Useful Lifetime (years)	Notes
Washer	500	-	11	10	Source [5,23,41]
Dryer	2790	-	6	13	Source [41–43]
Iron	1000	-	7	7	Source [5,23]
Microwave	1000	-	7	9	Source [5,23,41]
TV	138	-	35	7	Source [5,30]
PC	300	-	21	5	Source [5,30,43]
Refrigerator	-	907	-	13	Source [5,41]
Freezer	-	1037	-	11	Source [5,41]
Water cooler	-	799	-	10	Source [5], EERNGY STAR calculator (2.19 kWh/day)

3.3. Unit Power Demand

Unit power demand (*UPD*) is determined in a similar way to *UEC*, but only focuses on the equipment operating at the peak load period (i.e., summer in Kuwait) and can be expressed as [44]:

$$UPD = P \times RLF \times CDF \tag{10}$$

where P is the nameplate power per unit and rated load factor (*RLF*) is the ratio of the maximum operating demand of equipment to the rated input power. For example, air conditioners that operate above their rated input power could result in an *RLF* greater than one. The coincidence diversity factor (*CDF*) is used to account for the fact that not all stock units are operating at the peak time. The coincidence diversity factor is defined as the peak demand of a population of units at the system peak time to the peak demand of an individual unit, and can be expressed as [45]:

$$CDF = \frac{kW_{pop}}{\sum_{i=1}^{n}(kW_i \times RLF_i)} \tag{11}$$

where kW_{pop} is the peak demand of the population of units; kW_i is the nameplate rating of unit i; and RLF_i is the rated load factor of unit i.

3.4. Forecast Analysis

Figure 7 compares the predictions of the building energy stock model to the actual total energy consumption in Kuwait from 2005 to 2017 after a systematic calibration procedure. The actual energy consumption data were obtained from the Ministry of Electricity and Water (MEW) [2]. For the calibration analysis, three main input parameters were adjusted as follows:

1. The lighting power density was lowered to reflect the usage of energy efficient lighting fixtures in new and refurbished dwellings [39];
2. Archetypes C and D represent buildings implemented in the 1983 and 2010 MEW energy conservation codes. Therefore, the AC system COP (coefficient of performance) for Archetypes C and D was adjusted to be 2.4 and 2.9, respectively, to reflect the energy efficiency requirements by the MEW [46,47];
3. The household hot water usage was raised to 25 gallon per person per day to reflect the high per capita water consumption in Kuwait [48].

Good agreement between the predictions of the building energy stock model and the actual energy consumption was obtained with a relative error of less than 5%, as shown in Figure 7.

To predict a business-as-usual case (i.e., the baseline scenario), the forecast model mainly relies on the *UEC* and *stock*. With the projected values of *UEC* and *stock* for each equipment, we can use Equations (1) and (2) to predict the energy consumption and peak demand. The forecast of the equipment stock is mainly driven by the diffusion rate $D(y)$ and new housing construction. For Kuwait, the diffusion rate is higher due to the high levels of income and electrification rate.

In addition to the population data mentioned in the Introduction, Table 4 shows the amount of housing subsidies provided by the Public Authority of Housing Welfare (PAHW) each year that has been projected until 2034. The housing subsidy values are another driving variable used to estimate the stock included in the model by *First purchases(y)* in Equation (3).

Even with the base case scenarios, the efficiency of equipment and appliances tends to improve over the years. This was estimated by assigning an efficiency improvement rate for each equipment and appliance in the model. Depending on the equipment type, *UEC* was assumed to improve 1–5% in efficiency per year based on [31,32] to account for the technology changes and code requirements. In addition, some new technologies will diffuse into the market and replace old ones that can be less efficient. Light-emitting diode (LED) lighting is a good example since it was introduced in the Kuwaiti market a couple of years ago. A Bass model was used to estimate the adoption rate of LED lighting and was incorporated into the baseline model. The Bass model defines the fraction of sales $F(y)$ in year y to represent the adoption rate of a new technology or product as follows:

$$\frac{dF(y)}{dt} = (p + qF(y))(1 - F(y)) \tag{12}$$

where p represents the external factors that drive the market to adopt a new technology such as advertisement, and q is often referred to as the "word-of-mouth" effect from the early adopters to encourage the "imitators" to adopt the new technology [50]. To use the Bass model to forecast the adoption of a new product or technology, the parameter p (innovators), q (imitators), and the potential market size need to be estimated. Since no historical sales data of LEDs are available for Kuwait, the Bass model parameters were estimated by an analogy to the compact fluorescent lamps (CFL) that have past shipment data and similar diffusion characteristics with LEDs [51]. The ordinary least squares (OLS) method was used to estimate the Bass model parameters (i.e., the coefficient of innovation (p) and imitation (q)), as shown in Table 5.

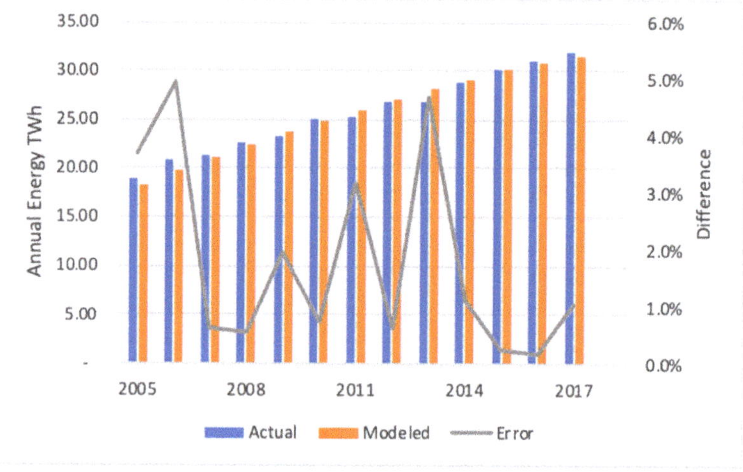

Figure 7. Comparison of the actual and modeled annual residential sector energy use from 2005 to 2017.

Several researchers have analyzed the effect of weather on energy consumption [52–54]. For the case of Kuwait, and based on [55–57], the influence of weather in the form of cooling degree-days (CDD) on long-term electricity demand forecasting is only statistically significant at 20% due to the low year-to-year weather variation in Kuwait. Therefore, the effect of annual weather variation was not considered in the forecasting model.

Table 4. Total housing subsidies provided by the Public Authority of Housing Welfare (PAHW) for various cities in Kuwait up to the year 2034 [49].

Area name	Area (m2)	2019	2020	2021	2022	2023	2024	2025	2026	2027	2028	2029	2030	2031	2032	2033	2034	Total
Al- Mutlaa	400		10,000	10,000	8288													28,288
West Abduallah Al- Mubarak	400	2000	2000	1201														5201
South Abduallah Al- Mubarak	400			1000	1260	1000												3260
South Saad Al- Abduallah	400				5000	5000	5000	5000	5000									25,000
Low- Cost Housing	200			2500	2500	2500	2500											10,000
Al Khairan	400						5000	5000	5000	10,000	5000	5000						35,000
South Sabah Al- Ahmed	400					5000	5000	5000	5000	5000	5000							30,000
Nawaf Al-Ahmed	600											10,000	10,000	5000	5000	5000	7000	42,000
Al Sabriya	400													10,000	10,000	11,000	11,000	42,000
Total		2000	12,000	14,701	17,048	13,500	17,500	15,000	15,000	15,000	10,000	15,000	10,000	15,000	15,000	16,000	18,000	

Table 5. Coefficients of innovation and imitation of the Bass model.

Parameter	p	q
Estimated value	0.0073	0.1686

4. Results and Discussion

Based on the specified inputs explained in the previous section, Figure 8 shows a bubble plot of the unit energy consumption of home appliances against the total stock to reveal the energy usage. The additional dimension, the size of the bubble, represents the total annual energy consumption. Household appliances included in the analysis consist of televisions (TV), personal computers (PC), washers, irons, microwave, refrigerator, freezer, water cooler, and dryers.

It can be observed that two major data clusters emerged with similar *UEC*s. One contained the following household appliances: televisions (TVs), personal computers (PCs), washers, irons, and microwaves. The *UEC* for this group ranged from approximately 200 kWh/year to 400 kWh/year. Despite the relatively low unitary electrical consumption of TVs, the quantity of the stock raised the level of impact. With approximately a thousand sets at a *UEC* of roughly 250 kWh per year, TVs represent a significant portion of the domestic energy use in Kuwait.

The second group consisting of higher *UEC*s, contained the following household appliances: water coolers, refrigerators, dryers, and freezers. Unlike refrigerators, freezers, and water coolers, dryers have low duty cycles and therefore consume less energy in a year, hence the smaller bubble. Moreover, in contrast, this group had a higher *UEC* range starting from approximately 800 kWh/yr to 1000 kWh/yr. Collectively, despite being less in stock, the overall impact is almost equally relevant due to the higher electrical consumption. This is partly due to the components that require significant power to operate such as compressors in refrigeration systems or resistive heaters commonly found in irons and electrical dryers.

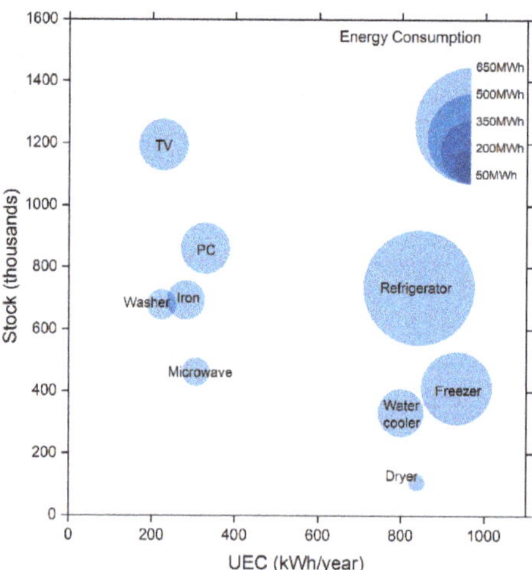

Figure 8. Electrical energy consumption of selected household appliances in Kuwait.

Table 6 displays the *UEC*, stock quantity, and the 2017 total energy consumption for specific household appliances for Kuwaiti homes. Due to the differences in energy use patterns between Kuwaiti and non-Kuwaiti homes, a similar analysis was conducted utilizing equal usage parameters, but with different stock quantities. The *UEC* values for the listed household appliances were calculated as outlined in the Methodology section and remain unchanged for both models. The highest *UEC* was noted to be freezers, refrigerators, and dryers, respectively, while PCs, washers, and TVs had *UEC*s that were less than a third that of freezers.

The results indicate that the energy consumption for the listed appliances totaled 1972.36 GWh for Kuwaiti households. Approximately 40% of the total consumption was attributed to refrigerators and freezers. The high energy consumption for these appliances was expected as the *UEC* values were high to begin with. However, due to the relatively high stock quantities, TVs also represented a significant load on the grid. Despite their low *UEC*s, the impact was offset by the volume, adding up to 926,505 TV sets, the highest stock quantity in all the listed appliances.

Table 6. The 2017 total energy consumption of the modeled household appliances with corresponding stock quantities for Kuwaiti residential homes (2017).

Plug Loads	*UEC* (kWh/year)	Stock	Total Energy Consumption (GWh)
Refrigerator	907.20	524,045	475.41
Freezer	1036.80	315,747	327.37
Washer	297.48	459,011	136.54
Dryer	882.57	89,540	79.03
Iron	375.95	470,321	176.82
Microwave	408.80	313,862	128.31
TV	251.85	926,505	233.34
PC	328.50	597,563	196.30
Water cooler	799.35	274,276	219.24
Total			1972.36

Utilizing the same list of appliances, along with their *UEC*, Table 7 shows the 2017 total energy consumption of modeled household appliances with corresponding stock quantities for expatriate (non-Kuwaiti) residential homes in Kuwait and displays the total energy consumption for specific household appliances for non-Kuwaiti homes.

Table 7. The 2017 total energy consumption of modeled household appliances with corresponding stock quantities for expatriate (non-Kuwaiti) residential homes in Kuwait.

Plug Loads	*UEC* (kWh/year)	Stock	Total Energy Consumption (GWh)
Refrigerator	907.20	211,349	191.74
Freezer	1036.80	96,724	100.28
Washer	297.48	221,084	65.77
Dryer	882.57	19,156	16.91
Iron	375.95	224,538	84.42
Microwave	408.80	150,111	61.37
TV	251.85	269,446	67.86
PC	328.50	264,107	86.76
Water cooler	799.35	58,097	46.44
Total			721.53

According to the results, the distribution of electricity consumption in residential households in Kuwait differs vastly, since the stock quantity weighs in heavily. Kuwaiti households account for roughly 70% of the total electrical consumption of the modeled appliances, whereas the remaining 30% was attributed to non-Kuwaiti household usage at 721.53 GWh. Parallel to the Kuwaiti profile, the results governing the expatriate households indicated that the top two energy-consuming appliances were refrigerators and freezers. The energy consumption of these two appliances make up approximately 40% of the overall energy usage for the expatriate household appliances.

From a broader perspective, the electrical consumption and demand distribution in residential households in Kuwait is broken down by the following main usage categories: lighting, air conditioning, space heating, water heating, and miscellaneous loads. Electrical consumption patterns remain heavily dependent on air-conditioning, as it represents the biggest slice within the pie charts shown in Figure 9. Air conditioning accounts for two thirds of the residential household consumption. However, although it is as little as 4.5%, space heating still accounts for a small load.

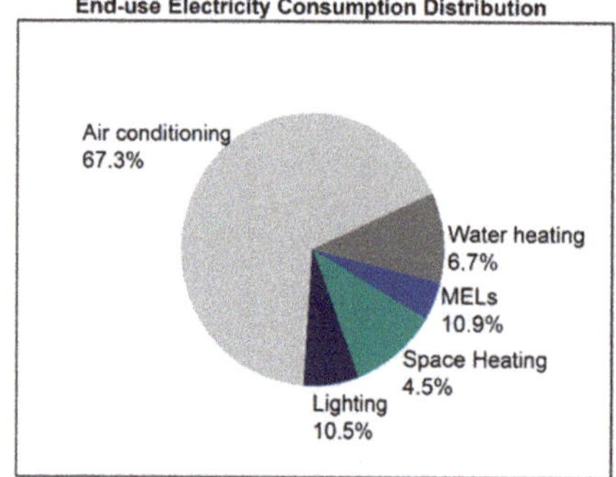

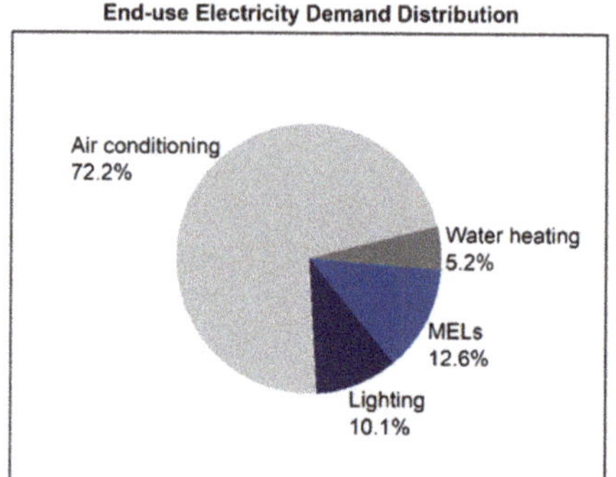

Figure 9. Distribution of on-site residential energy use and demand in Kuwait.

As seen in the distribution for residential electricity consumption, air conditioning makes up the bulk of the demand for Kuwaiti households at 66%, whereas the rest of the categories (miscellaneous loads, water heating, and lighting) range from 5% to 13%.

Utilizing the methodology outlined in this paper, the forecast of the residential energy consumption end-use was modeled and plotted in Figure 10. Revealing a similar trend observed in the household electrical consumption distribution and the household electricity demand distribution, air conditioning load is one of the highest loads for households. As per the results of the analysis, it is expected to rise exponentially from the year 2022 onward, reaching an estimated load of 60 TWh. Lighting is predicted to also rise, but much flatter, unlike the trend in air conditioning. The comparison between the actual and forecast points show an accurate model starting from 2005 until 2017. In terms of electrical demand, Figure 11 displays the growth for the air conditioning load, as it comprises a significant portion of the annual power demand. The results are also presented in a tabular form in Table 8. Figure 12 displays the forecast of electrical consumption for Kuwaiti and expatriate (non-Kuwaiti) households until the year 2040. Despite the slow growth in population, the forecast analysis indicates that the Kuwaiti energy consumption per capita was significantly higher than that of the expatriates, reaching levels of 15 MWh. The values for expatriates were almost stagnant, staying well below 1.5 MWh, despite the growing population figures that are expected to reach four million, more than doubling since 2005. The results are also represented in tabular form in Table 8.

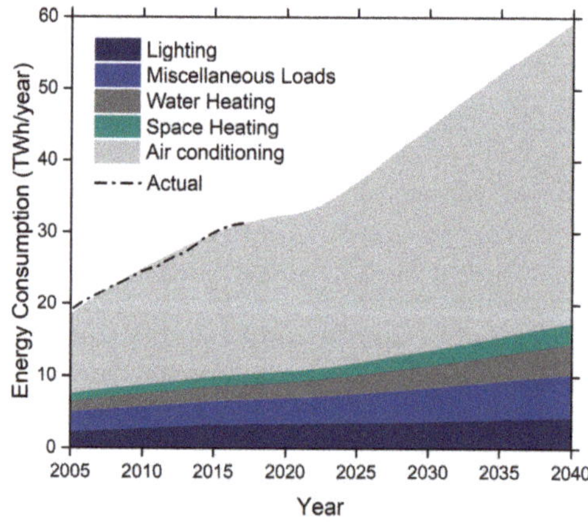

Figure 10. Forecast of on-site residential energy consumption by end-use until the year 2040.

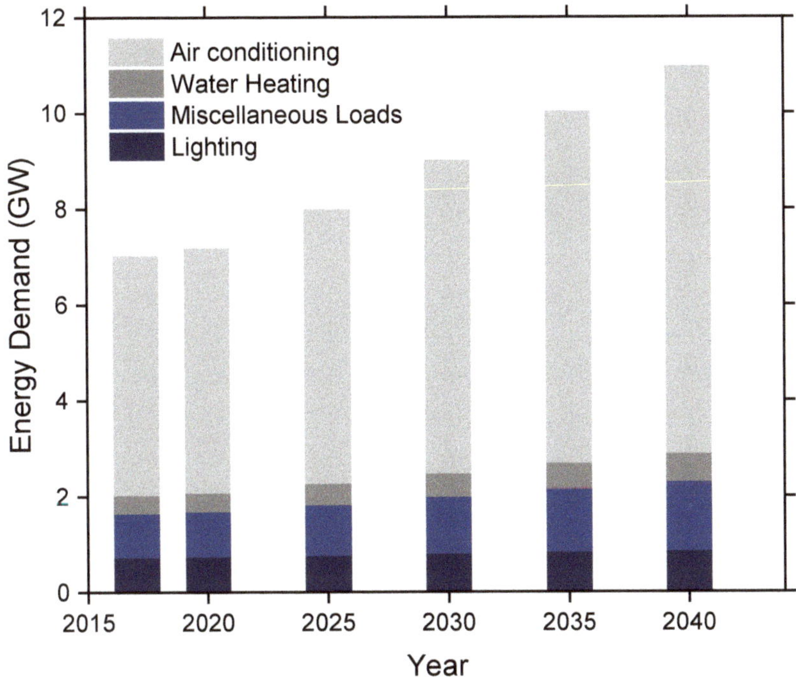

Figure 11. Forecast of on-site residential energy demand by end-use in Kuwait.

Table 8. Forecast of on-site residential energy consumption by end-use.

	Energy Consumption (GWh)					
	2017	**2020**	**2025**	**2030**	**2035**	**2040**
Miscellaneous Loads						
Kuwaiti	1687.61	1805.81	2100.36	2587.74	3095.81	3537.81
Expatriate	452.18	509.93	610.29	700.97	792.76	888.22
Total	2139.78	2315.74	2710.65	3288.71	3888.58	4426.04
Water Heating						
Kuwaiti	1821.13	1893.24	2244.82	2692.32	3184.99	3762.42
Expatriate	252.54	286.16	340.38	384.51	436.12	486.02
Total	2073.67	2179.40	2585.20	3076.83	3621.10	4248.44
Space Heating						
Kuwaiti	1672.00	1749.44	2026.52	2524.51	2987.72	3435.29
Expatriate	195.53	221.44	257.48	296.70	333.74	371.42
Total	1867.53	1970.88	2284.00	2821.21	3321.47	3806.71
Air Conditioning						
Kuwaiti	18,879.39	19,554.17	22,474.09	28,068.57	33,333.43	38,070.25
Expatriate	2027.28	2250.95	2697.75	3098.75	3486.88	3894.83
Total	20,906.67	21,805.12	25,171.84	31,167.32	36,820.31	41,965.08
Lighting	3527	3864	3866	3941	4246	4574
Total	30,514.51	32,135.44	36,617.78	44,295.47	51,897.46	59,020.41

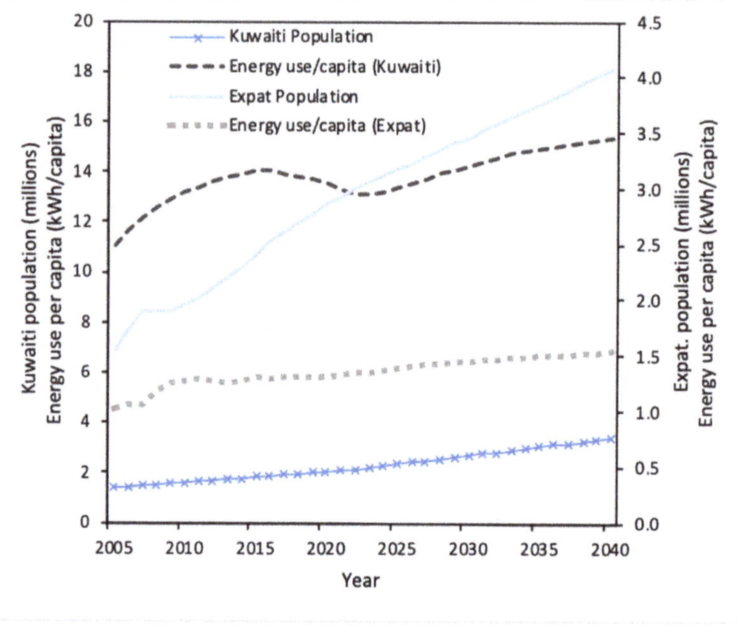

Figure 12. Forecast of population and residential energy consumption per capita.

5. Conclusions

Kuwait has one of the highest energy consumption per capita levels in the world. This large-scale consumption is negatively impacting its natural resources and the environment. The building sector alone accounts for 57% of electrical consumption. It is therefore important to study the driving impacts in a building's energy consumption in Kuwait. Utilizing end-use baseline information for residential loads sets an important foundation to help understand the residential consumption patterns. Based on the specified end-use equipment and certain parameters, a forecasting analysis was conducted to estimate the end-use distribution of electrical consumption for the state of Kuwait until the year 2040. In the model, end-uses were broken down into the following: air conditioning, lighting, miscellaneous loads, and space heating and water heating.

The resulting unit energy consumption (*UEC*) of home appliances was plotted against the total stock, which illustrated the impact of each of the specified home appliances. Refrigeration units, out of all appliances, held the highest *UEC* by far, as they were high in both stock and *UEC* values. A forecast model was then plotted to reveal the end-use energy consumption and peak demand in Kuwait until 2040. The air conditioning loads are expected to rise in the future with an average annual growth rate of 2.9%. Meanwhile, the rise in lighting energy consumption is much flatter due to an expected gradual shift toward more efficient lighting. Furthermore, based on the forecast results, differences between the Kuwaiti and expatriate (non-Kuwaiti) residential loads were observed. To the best of our knowledge, this is the first attempt to estimate the energy consumption of non-Kuwaiti households, where expatriates make up two-thirds of the population. These results provide opportunities for the development of more effective energy policies as well as opportunities for energy efficiency initiatives for the future.

The proposed model in this paper integrates equipment stock and unit energy consumption in order to project energy consumption at a more detailed level than other forecasting models. This level of detail in individual end-use equipment allows for the construction of various and detailed energy efficiency scenarios such as energy efficiency standards and labeling programs. Since the model accounts for replacement stock of equipment and appliances, this can also be used to evaluate energy retrofit programs. Moreover, this approach allows for data on equipment efficiency, sales, and stock over time to be separately developed, assessed, and incorporated into the model. The result is the ability to evaluate the stock turnover and penetration of energy-efficient equipment to the building stock, and their effect on the energy use and peak demand. It will also make the model more dynamic and updated based on the available sales data.

Author Contributions: Conceptualization, T.A. and P.P.; methodology, T.A.; software, T.A.; validation, T.A.; formal analysis, T.A. and P.P.; investigation, T.A. and P.P.; resources, T.A. and P.P.; data curation, T.A.; writing—original draft preparation, T.A.; writing—review and editing, P.P.; visualization, T.A.; supervision, P.P. All authors have read and agreed to the published version of the manuscript.

Funding: The research received no external funding.

Acknowledgments: The authors would like to express their gratitude to Barlas Demirciler, Nicholas Fette, and Ali Hajiah for their expertise and assistance throughout this study. The authors are also grateful to the Kuwait Ministry of Electricity and Water for their data and information support, especially Iqbal Al-Tayyar, Director of the Technical Supervision department.

Conflicts of Interest: The authors declare no conflict of interest.

References

1. World Bank. *World Development Indicators 2015*; World Bank: Washington, DC, USA, 2015; ISBN 9780821373866.
2. Ministry of Electricity and Water. *Statistical Year Book*; Ministry of Electricity and Water: Kuwait City, Kuwait, 2017.
3. US Energy Information Administration. *International Energy Statistics*; US Energy Information Administration: Washington, DC, USA, 2015.
4. Soares, N.; Reinhart, C.F.; Hajiah, A. Simulation-based analysis of the use of PCM-wallboards to reduce cooling energy demand and peak-loads in low-rise residential heavyweight buildings in Kuwait. *Build. Simul.* **2017**, *10*, 481–495. [CrossRef]
5. Ameer, B.; Krarti, M. Impact of subsidization on high energy performance designs for Kuwaiti residential buildings. *Energy Build.* **2016**, *116*, 249–262. [CrossRef]
6. Krarti, M.; Hajiah, A. Analysis of impact of daylight time savings on energy use of buildings in Kuwait. *Energy Policy* **2011**, *39*, 2319–2329. [CrossRef]
7. Wood, M.; Alsayegh, O. Electricity and Water Demand Behavior in Kuwait. In Proceedings of the 1st WSEAS International Conference on Natural Resource Management (NRM '12), Vienna, Austria, 10–12 November 2012; pp. 251–256.
8. Aydinalp-Koksal, M.; Ugursal, V.I. Comparison of neural network, conditional demand analysis, and engineering approaches for modeling end-use energy consumption in the residential sector. *Appl. Energy* **2008**, *85*, 271–296. [CrossRef]
9. Farahbakhsh, H.; Ugursal, V.I.; Fung, A.S. A residential end-use energy consumption model for Canada. *Int. J. Energy Res.* **1998**, *22*, 1133–1143. [CrossRef]
10. Tso, G.K.F.; Yau, K.K.W. Predicting electricity energy consumption: A comparison of regression analysis, decision tree and neural networks. *Energy* **2007**, *32*, 1761–1768. [CrossRef]
11. Wiesmann, D.; Lima Azevedo, I.; Ferrão, P.; Fernández, J.E. Residential electricity consumption in Portugal: Findings from top-down and bottom-up models. *Energy Policy* **2011**, *39*, 2772–2779. [CrossRef]
12. Zhao, H.X.; Magoulès, F. A review on the prediction of building energy consumption. *Renew. Sustain. Energy Rev.* **2012**, *16*, 3586–3592. [CrossRef]
13. Wilson, D.; Swisher, J. Exploring the gap. Top-down versus bottom-up analyses of the cost of mitigating global warming. *Energy Policy* **1993**, *21*, 249–263. [CrossRef]
14. Swan, L.G.; Ugursal, V.I. Modeling of end-use energy consumption in the residential sector: A review of modeling techniques. *Renew. Sustain. Energy Rev.* **2009**, *13*, 1819–1835. [CrossRef]
15. Nesbakken, R. Price sensitivity of residential energy consumption in Norway. *Energy Econ.* **1999**, *21*, 493–515. [CrossRef]

16. EIA. *Residential Demand Module of the National Energy Modeling System: Model Documentation*; US Energy Information Administration: Washington, DC, USA, 2014.

17. Kavgic, M.; Mavrogianni, A.; Mumovic, D.; Summerfield, A.; Stevanovic, Z.; Djurovic-Petrovic, M. A review of bottom-up building stock models for energy consumption in the residential sector. *Build. Environ.* **2010**, *45*, 1683–1697. [CrossRef]

18. Ghedamsi, R.; Settou, N.; Gouareh, A.; Khamouli, A.; Saifi, N.; Recioui, B.; Dokkar, B. Modeling and forecasting energy consumption for residential buildings in Algeria using bottom-up approach. *Energy Build.* **2016**, *121*, 309–317. [CrossRef]

19. Fumo, N.; Rafe Biswas, M.A. Regression analysis for prediction of residential energy consumption. *Renew. Sustain. Energy Rev.* **2015**, *47*, 332–343. [CrossRef]

20. International Energy Agency (IEA). *Energy Technology Perspectives 2017: Catalysing Energy Technology Transformations*; International Energy Agency: Paris, France, 2017; ISBN 978-92-64-27597-3.

21. Krarti, M. Evaluation of large scale building energy efficiency retrofit program in Kuwait. *Renew. Sustain. Energy Rev.* **2015**, *50*, 1069–1080. [CrossRef]

22. The Public Authority for Civil Information. *Statistical Reports*; The Public Authority for Civil Information: Kuwait City, Kuwait, 2017.

23. General Authority for Statistics. *Household Energy Survey*; General Authority for Statistics: Riyadh, Saudi Arabia, 2017.

24. Central Statistical Bureau. *Household Income and Expendeture Survey*; Central Statistical Bureau: Kuwait City, Kuwait, 2013.

25. McNeil, M.A.; Letschert, V.E. Modeling diffusion of electrical appliances in the residential sector. *Energy Build.* **2010**, *42*, 783–790. [CrossRef]

26. Zhou, N.; Fridley, D.; McNeil, M.; Zheng, N.; Letschert, V.; Ke, J.; Saheb, Y. Analysis of potential energy saving and CO_2 emission reduction of home appliances and commercial equipments in China. *Energy Policy* **2011**, *39*, 4541–4550. [CrossRef]

27. Radpour, S.; Hossain Mondal, M.A.; Kumar, A. Market penetration modeling of high energy efficiency appliances in the residential sector. *Energy* **2017**, *134*, 951–961. [CrossRef]

28. Bhattacharyya, S. International Handbook on the Energy Economics. *Int. J. Energy Sect. Manag.* **2010**, *4*, 482–486. [CrossRef]

29. Welch, C.; Rogers, B. Estimating the Remaining Useful Life of Residential Appliances. *ACEEE Summer Study Energy Effic. Build.* **2010**, *2*, 316–327.

30. Al-Mumin, A.; Khattab, O.; Sridhar, G. Occupants' behavior and activity patterns influencing the energy consumption in the Kuwaiti residences. *Energy Build.* **2003**, *35*, 549–559. [CrossRef]

31. EIA. *Updated Buildings Sector Appliance and Equipment Costs and Efficiencies*; US Energy Information Administration: Washington, DC, USA, 2016.

32. Natural Resources Canada. *Residential End-Use Model*; Natural Resources Canada: Ottawa, ON, Canada, 2017.

33. Shaban, N. *Development of typical meteorological year for Kuwait*; Report KISR 5857; Kuwait Institute for Scientific Research: Kuwait City, Kuwait, 2000.

34. Cerezo, C.; Sokol, J.; AlKhaled, S.; Reinhart, C.; Al-Mumin, A.; Hajiah, A. Comparison of four building archetype characterization methods in urban building energy modeling (UBEM): A residential case study in Kuwait City. *Energy Build.* **2017**, *154*, 321–334. [CrossRef]

35. Almutairi, H. Low Energy Air Conditioning for Hot Climates. Ph.D. Thesis, University of Manchester, Manchester, UK, 2012.

36. De Wolf, C.; Cerezo, C.; Murtadhawi, Z.; Hajiah, A.; Al Mumin, A.; Ochsendorf, J.; Reinhart, C. Life cycle building impact of a Middle Eastern residential neighborhood. *Energy* **2017**, *134*, 336–348. [CrossRef]

37. Aguilar, C.; White, D.J.; Ryan, D.L. *Domestic water heating and water heater energy consumption in Canada*; CBEEDAC: Edmonton, AB, Canada, 2005.

38. U.S. Department of Energy. *ENERGY STAR Water Heater Market Profile*; U.S. Department of Energy: Washington, DC, USA, 2010.

39. UNEP. *Country Lighting Assessment*; UNEP: Paris, France, 2010.

40. Gelil, I.A. *Regional Report on Efficient Lighting in the Middle East and North Africa*; United Nations Environment Programme: Nairobi, Kenya, 2011.

41. Seiders, D.; Ahluwalia, G.; Melman, S. *Study of life expectancy of home components*; National Association of Home Builders, Bank of America Home Equity: Washington, DC, USA, 2007.
42. DOE Energy Saver for Appliances and Electronics. Available online: http://energy.gov/energysaver/articles/estimating-appliance-and-home-electronic-energy-use (accessed on 1 November 2018).
43. Size, A. Appliance Usage. Available online: https://www.aps.com/en/residential/savemoneyandenergy/homeappliance/Pages/appliance-usage.aspx (accessed on 1 November 2018).
44. Sturn, F.; Spencer, J. *The Uniform Methods Project: Methods for Determining Energy Efficiency Savings for Specific Measures*; Office of Energy Efficiency & Renewable Energy: Washington, DC, USA, 2016.
45. CPUC. *The California Evaluation Framework*; CPUC: San Francisco, CA, USA, 2004.
46. Debs, A.S. Energy conservation in Kuwaiti buildings. In *Energy Conservation Measures*; Elsevier: Amsterdam, The Netherlands, 1984; pp. 85–106.
47. Ministry of Electricity and Water. *Energy Conservation Code of Practice, MEW/R-6/2010*; Ministry of Electricity and Water: Kuwait City, Kuwait, 2010.
48. AlMayyas, H.; Leeb, S. *Modelling Kuwait Water System Using Simulink*; Department of Electrical Engineering Massachusetts Institute of Technology: Cambridge, MA, USA, 2015.
49. Public Authority for Housing Welfare, Kuwait City, Kuwait. Available online: http://www.housing.gov.kw (accessed on 1 May 2018).
50. Bass, F.M. A New Product Growth for Model Consumer Durables. *Manag. Sci.* **1969**, *15*, 215–227. [CrossRef]
51. Lilien, G.L.; Rangaswamy, A.; Van den Bulte, C. *Diffusion models: Managerial applications and software*; New product diffusion models; Kluwer Academic Publishers: Boston, MA, USA, 2000.
52. Considine, T.J. The impacts of weather variations on energy demand and carbon emissions. *Resour. Energy Econ.* **2000**, *22*, 295–314. [CrossRef]
53. Elkhafif, M.A.T. An iterative approach for weather-correcting energy consumption data. *Energy Econ.* **1996**, *18*, 221–230. [CrossRef]
54. Olonscheck, M.; Holsten, A.; Kropp, J.P. Heating and cooling energy demand and related emissions of the German residential building stock under climate change. *Energy Policy* **2011**, *39*, 4795–4806. [CrossRef]
55. Alarenan, S.; Gasim, A.A.; Hunt, L.C.; Muhsen, A.R. Measuring underlying energy efficiency in the GCC countries using a newly constructed dataset. *Energy Transit.* **2019**, *3*, 31–44. [CrossRef]
56. Atalla, T.N.; Hunt, L.C. Modelling residential electricity demand in the GCC countries. *Energy Econ.* **2016**, *59*, 149–158. [CrossRef]
57. Atalla, T.; Gualdi, S.; Lanza, A. A global degree days database for energy-related applications. *Energy* **2018**, *143*, 1048–1055. [CrossRef]

© 2020 by the authors. Licensee MDPI, Basel, Switzerland. This article is an open access article distributed under the terms and conditions of the Creative Commons Attribution (CC BY) license (http://creativecommons.org/licenses/by/4.0/).

Article

Eco-Efficient Value Creation of Residential Street Lighting Systems by Simultaneously Analysing the Value, the Costs and the Eco-Costs during the Design and Engineering Phase

Nine Klaassen, Arno Scheepens, Bas Flipsen and Joost Vogtlander *

Department Design Engineering, Delft University of Technology, Landbergstraat 15,
2628 CE Delft, The Netherlands; nine.klaassen@gmail.com (N.K.); A.E.Scheepens@tudelft.nl (A.S.);
S.F.J.Flipsen@tudelft.nl (B.F.)
* Correspondence: j.g.vogtlander@tudelft.nl

Received: 13 May 2020; Accepted: 28 June 2020; Published: 30 June 2020

Abstract: In search of sustainable business models, product innovation must fulfil a double objective: the new product must have a higher (market) value, and at the same time a lower eco-burden. To achieve this objective, it is an imperative that the value, the total costs of ownership, and the eco-burden of a product are analysed at the beginning of the design process (idea generation and concept development). The design approach that supports such a design objective, is called Eco-efficient Value Creation (EVC). This approach is characterised by a two-dimensional representation: the eco-burden at the y-axis and the costs or the value at the x-axis. The value is either the Willingness to Pay or the market price. The eco-burden is expressed in eco-costs, a monetised single indicator in LCA (Life Cycle Assessment): an app for IOS and Android, and excel look-up tables at the internet, enable quick assessment of eco-costs. A practical example is given: the design of a new concept of domestic street lighting system for the city of Rotterdam. This new concept results in a considerable reduction of carbon footprint and eco-costs, and shows the benefits for the municipality and for the residents, resulting in a viable business case.

Keywords: street lighting system; TCO; EVR; EVC; eco-efficient value creation; eco-costs

---

1. Introduction

1.1. The Issue: Progress in Sustainable Product Innovation, and Circular Business Models

There is a general concern about the increasing concentration of greenhouse gases in our atmosphere, materials scarcity, degradation of biodiversity, the plastic soup in the oceans, and many other pollutants like fine dust and NOx. As politicians set stricter targets (Kyoto protocol, Paris Agreement), and citizens become more and more aware of the severe consequences, business people realise that they should innovate their products and services. New business proposals must have a double objective: the new product must have a higher customer value, and at the same time a lower eco-burden. The higher customer value is needed to make the introduction of the product at the market a success, without the need for state subsidies.

The fact is that sustainable product innovation and introduction of circular business models is not easy. Although circular business models became a hype in Western Europe after the introduction of the Cradle-to-Cradle philosophy [1], and the business aspects of it [2,3], real successful implementations are rather limited [4–7] for many commercial business reasons. In addition, the environmental gains of circular business models are often much lower than it is suggested, especially with regard to the shift to services [8]. White et al. [9] (p. 1) writes about services: *"It is clear that the simplest and most optimistic view—a service economy is inherently clean economy—is insufficient and incorrect. Instead, the service economy is better characterized as a value-added layer resting upon a material-intensive, industrial economy"*. Tukker [10] has drawn a similar conclusion on Product Service Systems (PSS) after the comprehensive SusProNet study: PSS did not bring the enormous change that was hoped for. PSS might *support* new business models, but are not the solution as such. White et al. conclude that, even though growth in services might be less environmentally damaging than growth in manufacturing: *"If services are to produce a greener economy, it will be because they change the ways in which products are made, used and disposed of—or because services, in some cases, supplant products altogether"* [8] (p. 1). Therefore, one of the crucial aspects of the innovative design of sustainable products or services is that people will buy it and use it (instead of unsustainable alternatives). That is why we focus in this paper on value creation in the fuzzy front end of the design process, since that is the moment where the real sustainable innovation can take place. It is the moment for designers to contemplate radically different product or service systems, e.g., identifying 'functional result' alternatives [9]. In user-centred design, value creation for the customer is the main aim [11]. In Ecodesign, sustainability is the main aim [12]. However, in sustainable product innovation we need a combination of both [13], where value creation and sustainability go hand in hand, and where the classical contradiction between ecology and economy is being reconciled in a clever way. Only LCA (Life Cycle Assessment) can reveal to what extent a new design of a product chain is better in terms of sustainability. A practical issue in design is that the classical LCA method is too laborious and complex to be doable in the early design stages. The result is that, especially in the early design stages, the aspect of sustainability is only dealt with on a qualitative 'gut feeling' basis, often leading to wrong conclusions. The LCA is then done (if at all) when the detailed design is ready. At that stage, however, it is too late for drastic changes, resulting in a product design that is far from the optimum. In eco-efficient value creation, this issue has been solved in a practical way by tools for 'Fast Track LCA', enabling the assessment of the environmental impacts of multiple design concepts in a quick way.

1.2. The Challenge: A Sustainable Street Lighting System for the City of Rotterdam

This paper presents the results of a practical case of eco-efficient value creation. It is the design of a street lighting system for a typical city in Western Europe: the city of Rotterdam.

Public street lighting has a major influence on safety [14,15], the perception of safety, and in general the atmosphere in the street. The municipality of Rotterdam has the desire to create a pleasant atmosphere in the city during both day and night by street lighting systems. An additional aspect of well-being in cities is the local presence of nature, i.e., trees [16,17]. So lighting and trees are both important aspects of the value for the citizens. However, in the conventional design of street lighting systems, there is a conflict below the ground: the roots of the trees interfere with the power cables, see Figure 1. This conflict causes difficulties during installation, maintenance and operation as well as end-of-life, which all lead to higher costs for the application of street lighting systems.

The underground conflict between tree roots and power cables can be solved in two different ways: (1) find other ways to create residential green, e.g., with plant boxes; or (2) redesign public street lighting. Within the first direction many solutions can be found, however, that is not the scope of this paper. To find acceptable solutions for the second direction is not easy. Since 1800, the lamppost has looked the same: a light source on a pole. Other forms such as hanging street lighting with hanging power cables above the ground are generally not regarded as desirable.

From the point of view of sustainability, the system requirement is obvious: the design must combine LED lighting with local PV cells as the source for the required electricity. Replacing the classic lamps by LED lamps is easy. Where to place the PV cells is less easy: (1) PV cells above the street will require expensive construction; (2) PV cells on the roof are a logical choice, but why would the owner of the building allow the municipality to attach the PV cells?

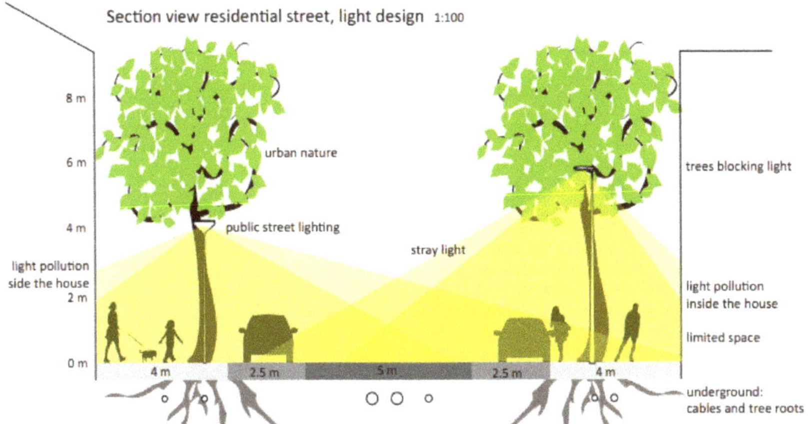

Figure 1. Conflict between residential street lighting systems and trees: the power cables are entangled in the roots.

The design of a new street lighting system has to fulfil three value aspects for the 3 stakeholders: (1) the requirement of streetlights in combination with trees, which is the value for the citizens; (2) it must be affordable (not too expensive) for the municipality; (3) it must resolve the issue "what is in it for me?" for the house owner with regard to the PV cells. At the same time, the new system must have a (much) lower eco-burden score in LCA compared to the classical system of Figure 1.

2. The Methods

2.1. The Eco-Costs, a Monetized Single Indicator in LCA

The assessment of the eco-burden of a system is done by LCA. An important issue here is the choice of the indicator that is used for benchmarking. Such a benchmarking indicator can be a so called midpoint indicator (e.g., greenhouse gas, acidification, eutrification, fine dust, human toxicity, ecotoxicity), but the issue here is that every indicator leads to its own optimum choice in product design. A well-known example is the engineering of the Volkswagen diesel: by focusing on CO_2 emissions only, and ignoring the consequences for NOx emissions, the strategic decisions of the company lead to losses of several billion euros.

The solution is to apply a so called endpoint indicator, which combines all midpoint indicators in one single score (i.e., damage based indicators like ReCiPe [18] and Ecological Footprint [19], both in 'points', or monetized scores like EPS [20] and eco-costs [21]). There is no single truth in single endpoint indicator systems, since such a system reflects a set of values and assumptions, but it is generally acknowledged that single score systems are needed in LCA benchmarking. A well-documented scientific single indicator system is always better than a set of many midpoint scores of which one or two are selected on the basis of a personal, subjective point of view [22,23].

It is useful to select a monetised single indicator in LCA, since it is related to the concept of 'external costs' (i.e., environmental costs to our society that are not included in the current product costs) and thus enables the comparison with the costs and the market value of the design. In the scientific literature there are two operational monetized systems that are widely applied in LCA: EPS 2015 (a damage-based indicator) [20] and Eco-costs 2017 (a prevention-based indicator) [21]. The advantage of monetized systems is that they do not suffer from the inaccuracies of the normalisation and weighting steps.

For the street lighting system study in Rotterdam, the eco-costs was selected as a monetised single indicator, since it is the most comprehensive system in terms of midpoints, see Figure 2, and it is the most applied system in science as well as design engineering.

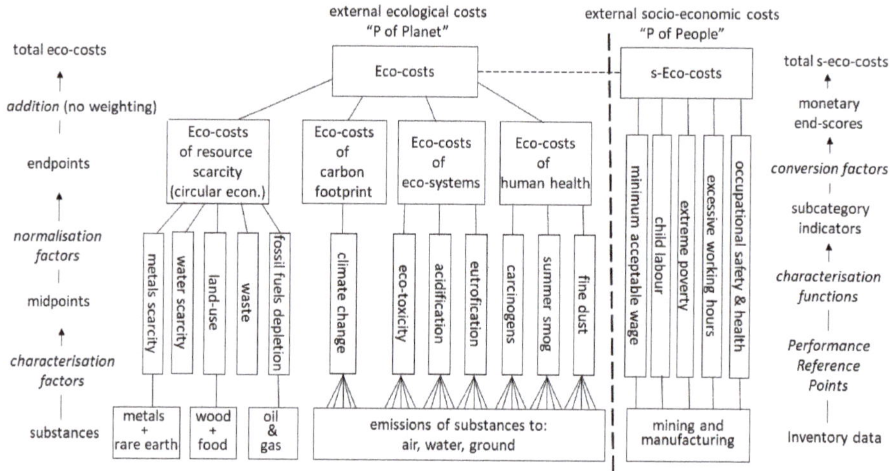

Figure 2. The total eco-cost system in life-cycle assessment.

The eco-costs system has been developed in the period 1999–2002 [24–26] and updated in 2007, 2012 and 2017 [27]. The system is in compliance with ISO 14008 [28]. A further description of the monetisation factors can be found in [29].

The way the total eco-costs of a system like street lighting are calculated, is explained by Figure 3. The first step in LCA is to determine the so called Life Cycle Inventory (LCI) list of all polluting emissions (CO_2, SO_2, NOx, fine dust, etcetera) and all required resources (metals, energy carriers, water, land). The system delivers a product or service as output (in this case light), and comprises a lot of subsystems and processes (in this case the lampposts, the cables, the light bulbs, the installation processes, and the end-of-life processes). All these subsystems and processes need material, transport and energy (electricity and heat) as input.

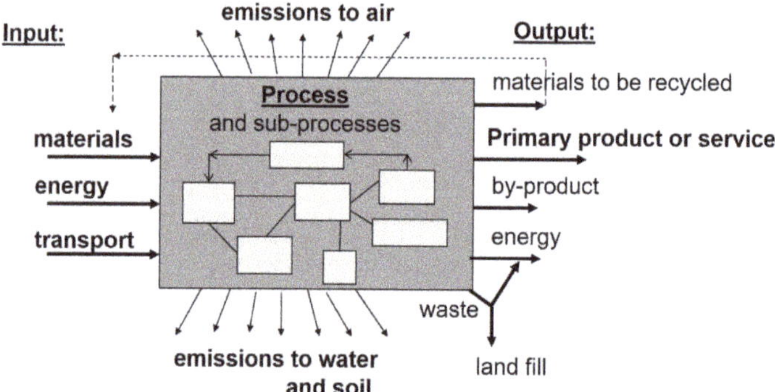

Figure 3. The system components of Life Cycle Assessment.

The second step in LCA is called the Life Cycle Impact Assessment (LCIA). The goal of this step is to provide a practical interpretation of the long list of emissions and required resources of Figure 3. According to ISO 14044 [30], this is done via the calculation structure of Figure 2. The substances of the list are classified in terms of their effect, multiplied by characterisation factors, and added up within their own 'midpoint' groups (i.e., climate change, eco-toxicity, acidification, fine dust, carcinogens, etcetera). Then the midpoint groups are combined to 'endpoints' (so called Areas of Protection) after either a monetisation step (e.g., eco-costs), or by 'normalisation' (e.g., 'points' in the ReCiPe system).

In the case of monetisation, the 'endpoints' can be added up to a total end-score, in our case eco-costs. (Non-monetised systems need an extra step to weight the relative importance of the points of the Areas of Protection).

LCA calculations can be made either with special software (e.g., Simapro, Gabi, Open LCA), or by means of look-up tables in excel. These tables are available for eco-costs of pure emissions, but also for the aggregated eco-costs at the level of materials (metals, plastics, wood etc.), manufacturing processes (deep drawing, turning, welding, extrusion, coating etc.), components (lamp bulbs, printed circuit boards, PV panels), transport, energy, and end-of-life processes [31]. These look-up tables have been calculated with the use of formal LCI databases, and enable a simplification of the final LCA calculation (without losing accuracy) in a way that is quite similar to cost accounting in projects (multiplying quantities with its eco-costs scores of supplies and processes, and adding it up to the total eco-costs). An example of such an LCA is given in Table 1. The table provides output data (in eco-costs and in CO_2 equivalent) for one classical lamppost (type 'Kegeltop' on a 4 m pole). Note that the calculations in Section 3 (Results) show data per year, under the assumption that the lifespan of a lamppost is 40 years, and per street, under the assumption that a street has 100 lampposts.

Table 1. An example of an excel LCA table, based on the BOM (bulk of materials) of the base case of a classical lamppost in a street (type 'Kegeltop' on a 4 m pole, per lamppost).

Unit	Amount for 1 FU	LCI Database Line	Eco-Costs Per Unit	CO2e Per Unit	Eco-Costs Per FU	CO2e Per FU
kg	3.52	Aluminium trade mix (45% prim 55% sec)	2.12	6.26	7.5	22.0
kg	34.93	Steel (21% sec = market mix average)	0.60	1.61	20.8	56.2
kg	2.10	Polyester (unsaturated) 70%	2.04	7.46	4.3	15.7
kg	0.90	Glass fibre 30%	0.10	0.48	0.1	0.4
kg	1.78	PC pellets	2.05	7.78	3.7	13.9
kg	0.07	Copper trade mix (56% prim 44% sec)	2.70	1.82	0.2	0.1
kg	0.31	PP pellets	1.05	1.97	0.3	0.6
kg	0.04	ABS pellets 50%	1.32	3.40	0.1	0.1
kg	0.04	PC pellets 50%	2.05	7.78	0.1	0.3
kg	1.50	Glass, uncoated	0.22	0.98	0.3	1.5
kg	0.08	PWB desktop, including components and Ics	60.71	160.41	4.9	12.8
m	16.60	Electric cord, 1000 W, 3×0.5 mm^2, domestic	0.07	0.14	1.2	2.3
kg	0.06	Crude iron	0.42	1.51	0.0	0.1
kg	0.002	Silicon	2.21	10.59	0.0	0.0
kg	0.07	67SiCr5, spring-steel	0.77	1.85	0.1	0.1
kg	0.55	X5CrNi18 (Stainless steel 304)	2.66	3.85	1.5	2.1
kg	0.41	PVC	0.702	2.006	0.3	0.8
		Material supplies			45	129
		Production processes				
kg	34.00	Drawing of pipe, steel	0.170	0.360	5.8	12.2
m2	10.25	Electroplating Zinc, outside use, per 10 years	5.479	2.974	56.2	30.5
m2	2.56	Powder coating, steel/RER S	1.105	4.570	2.8	11.7
kg	5.43	Injection molding plastics	0.264	1.333	1.4	7.2
kg	2.02	Casting, aluminium	0.018	0.157	0.0	0.3
kg	1.50	Cold transforming Al	0.019	0.104	0.0	0.2
kg	0.99	Deep drawing steel	0.065	0.316	0.1	0.3
		Manufacturing processes			66	62
		Material supplies + manufacturing processes excluding transport and installation			112	192

2.2. The Model of the Eco-Costs/Value Ratio

The basic idea of the model of the Eco-costs / Value Ratio (EVR) is to link the value chain of Porter [32], to the ecological product chain. In the value chain, the added value (in terms of money) and the added costs (from Life Cycle Costing, LCC) are determined for each step of the product chain, cradle-to-grave. Similarly, the ecological impact of each step in the product chain is expressed in terms of money, the eco-costs. See Figure 4.

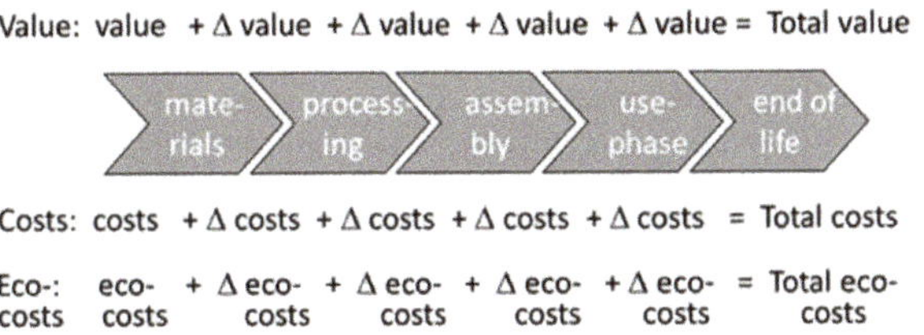

Figure 4. The basic idea of the Eco-costs/Value Ratio (EVR): combining the value chain with the ecological chain [33].

The theory of Porter, and so Figure 4, deals with the manufacturing of (physical) products for end-users (consumers). In a slightly more complex form, this theory can also describe the 'profit pool' [34] of a circular business model, or a service, since industrial services are bundles of products that deliver a function to the end-user. Street lighting is an example of such a service: its main function is light at night to provide safety, delivered by a bundle of products and services (lampposts, electricity, and maintenance). It is important here to realise that the value (of a product or service) for an individual buyer is not equal to the market price. The value is the Customer Perceived Value (CPV) [35–37], also called Willingness to Pay. The relationship between the costs, the price and the CPV is depicted in Figure 5.

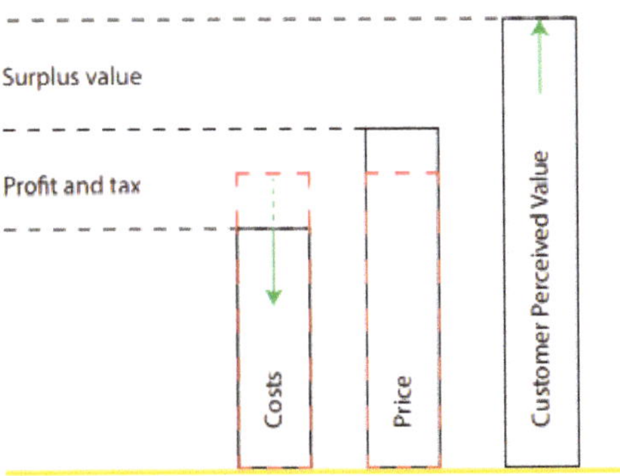

Figure 5. The costs, the price, and the Customer Perceived Value (CPV) of a product.

In our free market economy, the costs should be lower than the price, to support the profit of the company without subsidies. On the other hand, a product can only be marketed successfully when the CPV is higher than the market price, since people tend to buy things only when the perceived value for them is higher than the price they have to pay. The CPV can be defined as the benefit (utility plus joy) that is expected after the purchase. We call the difference between the price and the CPV the Surplus Value for the individual buyer. In the free market economy, the (market) price is set at a level that attracts sufficient buyers in order to reach an economy of scale that keeps the costs low enough.

For a municipality, costs and price are the same (red dotted lines in Figure 5), because they do not have the goal of making profit. However, the surplus value (for its citizens) must be positive, otherwise a project will not be accepted by the public.

In fact, the EVR model entails multiple dimensions. However, to show the build-up of the product in the chain, it is better in most cases to display only two dimensions at a time (see the figures in Sections 3.1 and 3.2 as an example for the base case of streetlighting) to avoid complex 3-D charts: the eco-costs at the y-axis, and one of the financial dimensions at the x-axis.

Under the assumption that most of the households spend in their life what they earn in their life (the bank savings ratio is <5% in most countries), the total EVR of the spending of households is the key towards sustainability. Only when this total EVR of the spending is consistently lowered, the eco-costs related to the total spending will be reduced (even at a higher level of spending). This issue is explained by a short macro-economic analysis on what happens in the European Union. Figure 6 shows the EVR (= eco-costs/price) on the Y-axis as a function of the cumulative expenditures of all products and services of all citizens in the EU25 on the X-axis. The data is derived from the EIPRO study of the European Commission (EIPRO = environmental impact of products) [38].

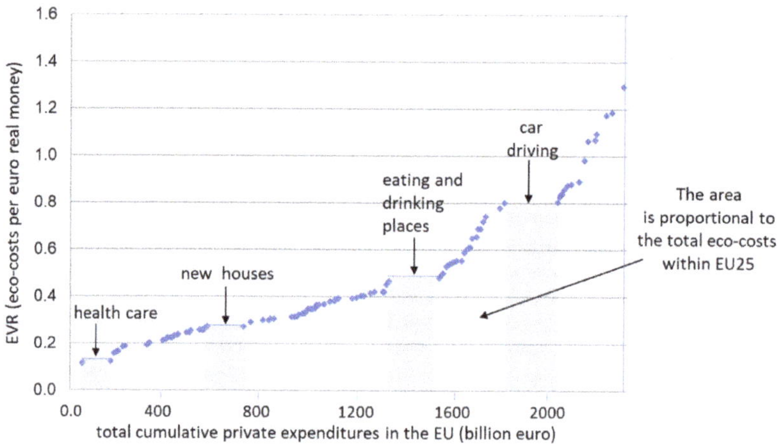

Figure 6. The EVR and the total expenditures of all consumers in the EU25 (from the environmental impact of products (EIPRO) study [38]).

The area underneath the curve is proportional to the total eco-costs of the EU25. Basically, there are two strategies to reduce the area under the curve:

- force industry to reduce the eco-costs of their products (this will shift the curve downward);
- try to reduce expenditures of consumers in the high end of the curve, by attractive offerings at the low end of the curve (this will shift the middle part of the curve to the right).

The question is now how designers and engineers can contribute to this required shift towards sustainability. Key is product innovation that fulfils the double objective of a higher CPV, and at the same time a lower eco-burden. To achieve this objective, it is an imperative that the designer must look at the CPV as well as the eco-costs at the beginning of the design process (i.e., idea generation and concept development). Eco-efficient Value Creation is a structured design method to achieve this.

2.3. Eco-Efficient Value Creation

In search of sustainable business models, product innovation must fulfil the double objective of eco-efficiency [39–41]. To achieve this objective, it is an imperative that the value, the total costs of ownership, and the eco-burden of a product are analysed at the beginning of the design process (idea generation and concept development).

The successful design options for Eco-efficient Value Creation are:

- to increase value where value is high (more quality, service, life span, and image);
- to decrease the eco-costs where the eco-costs are high (a shift to bio-based materials, recycling and renewable energy).

End-of-life solutions are important as well. Landfill reduces the value of the total system, and leads to higher eco-costs. Recycling (as well as re-use and remanufacturing) results in an added value combined with lower eco-costs ('end-of-life credits' in LCA).

A comprehensive checklist on the reduction of eco-costs is provided by the LiDS Wheel of Eco-Design [12], but the real issue of eco-efficient value creation is how to enhance the Customer Perceived Value of a green product at the same time. Mestre [13] studied the eco-efficient value creation with cork as bio-based material, and described the basic principles for the fuzzy front end of the design, see Figure 7, where, according to Mestre, "it is the talent of the designer that creates the value of the product" (page 13). In fact, sometimes a bit more eco-costs must be allowed to enhance the value considerably, leading to a better EVR score of the design.

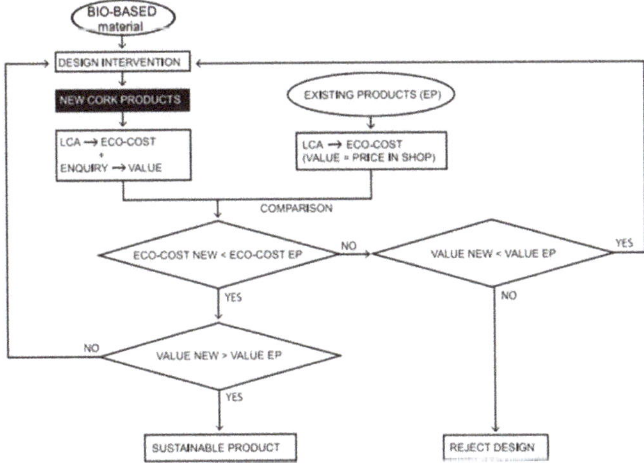

Figure 7. The basics of eco-efficient value creation in the fuzzy front end of the design [13].

Figure 7 clearly shows that the transformation towards a circular economy fulfils the double obligation of eco-efficient value creation. However, it also shows that designing a sustainable circular system needs to address more than circularity only: other aspects such as clean production, minimum transport and optimal marketing play an important role as well. To assess the environmental aspects (eco-costs), LCA is an indispensable tool throughout all stages of product development, see Figure 8. However, the classical LCA approach is only doable at the final detailed design stage, because it is too laborious [42]. To enable LCA-based materials selection in the fuzzy front end of idea generation, excel look-up tables [31] and an app for IOS and Android have been developed [43]. A special version of this app can make Fast Track LCAs, to optimize the design in the concept development phase (e.g., to analyse the trade-off of choices on materials, transport distances, and required energies).

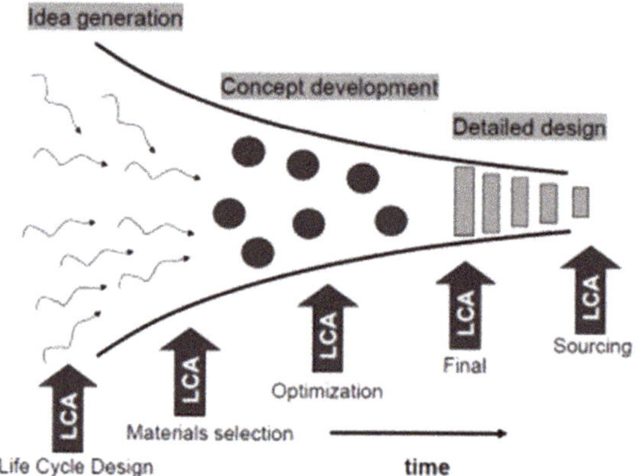

Figure 8. The use of LCA during all product development stages [44].

The approach of eco-efficient value creation can be characterised by 6 sequential steps:

Stpe 1 Life Cycle Thinking: At the start of the design process, the basic questions on circular design are whether or not the product must be suitable for easy repair, takeback + remanufacturing, or takeback + recycling of the materials. Note that circular designs are not always realistic in practice (because of long life times, high costs of return transport to the factory, low quantities, high remanufacturing costs, governmental regulations etc.). So Life Cycle Thinking must comprise many aspects that are on a higher level than the product chain itself [45].

Step 2 Functional requirements, and possible add-ons to enhance the CPV: Establish the 'musts' and the 'wants' in terms of functionalities, and in terms of enhancing the CPV [46].

Step 3 Idea generation and materials selection: The designer might be inspired by biomimicry, nature-inspired design, bio-inspired design, C2C, and other philosophies and design tools [47]. Since the choice of materials plays a governing role in this design stage [48], the LCA-based Idemat app for materials selection (specially developed to support eco-efficient value design) might be applied [43].

Step 4 Concept development and design optimisation: This is a highly iterative process as depicted in Figure 7.

Step 5 Detailed design with a final product LCA and with sourcing of components (materials): This is the stage of the classical LCA, to find the environmental the hotspots of the final design.

Step 6 Selection of suppliers: At the stage of sourcing of the components and materials, LCA should be applied to select the preferred suppliers.

3. Results: Example of the Design of a Street Lighting System

3.1. Base Case: the EVR of a Traditional Design in the City of Rotterdam

The base case for the design is the currently dominant existing system. The chosen lamppost for this base case is the "Kegeltop", on a 4 m aluminium pole. This luminaire is one of the most used ones in residential streets in the Netherlands and is a well-known design.

The Functional Unit (FU) of the analysis is: (1) one street, 1200 × 20 m (2) one year with a light level according to regulations (minimum of 3 lux at street level and a uniformity rate of at least 25%). The life span of a lamp post system is set to 40 years. The life span of PV cells is assumed to be 20 years.

The Total Costs of Ownership (TCO) of the base case comprises:

- Manufacturing and installation costs. These costs include the purchasing costs of the pole and luminaire and the working hours and administration costs of the installation process. Creating a grid connection, digging for cables and the pole are expensive: about 55% of the installing costs. Purchasing the pole and luminaire is the other 45%.

- Technical management. This is mainly related to maintenance work, such as: replace light bulbs, repair electronics and cable failures (after accidents), clean luminaires.
- Administrative management. These costs are related to desk work. Examples are: office expenses and taxes, inspections of luminaries and processing of the inspection reports.
- Energy consumption. This is based on the most used light source for residential streets: 36 W PL fluorescent lamps. The yearly operating time of a single light bulb is 4200 h. In addition, some taxes are included in the energy consumption costs.
- End of life. These are the costs for the removal tax of a pole and luminaire, and the removal costs of the current grid connection.

The TCO of the base case is depicted in Figure 9.

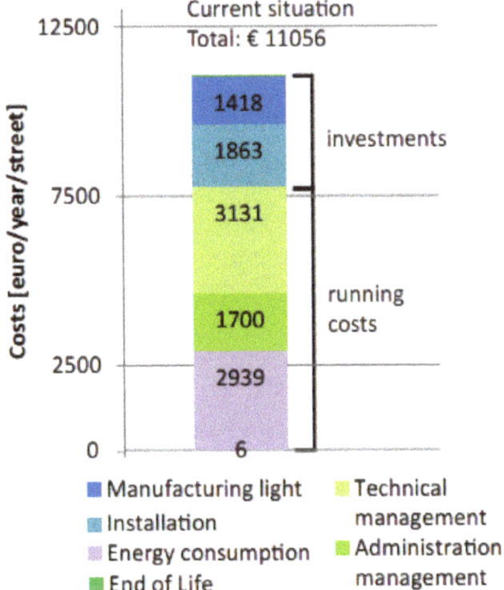

Figure 9. Costs (Total Costs of Ownership, TCO) for lighting: one street for one year of the base case.

The eco-costs of lighting system is depicted in Figure 10. Not all issues of the TCO have relevant eco-costs: administration and technical management consist out of labour, which is usually not part of an LCA. Maintenance does require some car kilometres to be driven, but that can be neglected in the LCA. The eco-costs of the energy consumption are highest together with the eco-costs of manufacturing. The eco-costs of the End of Life phase are negative since the material of the pole (aluminium) is reused in the circular business model of the contractor.

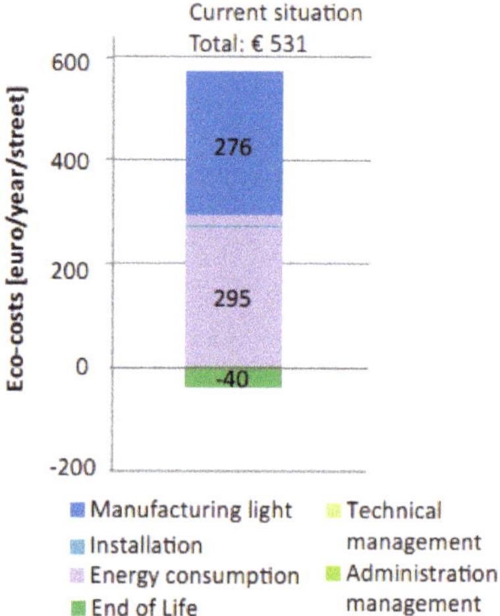

Figure 10. Eco-costs costs for lighting: one street for one year of the base case.

The costs, which represent the value, and eco-costs are plotted against each other in Figure 11. This graph shows which life cycle steps are most harmful for the environment and which steps are most expensive. The EVR ratio is highest after the manufacturing phase followed by the energy consumption, installation phase and technical and administration management.

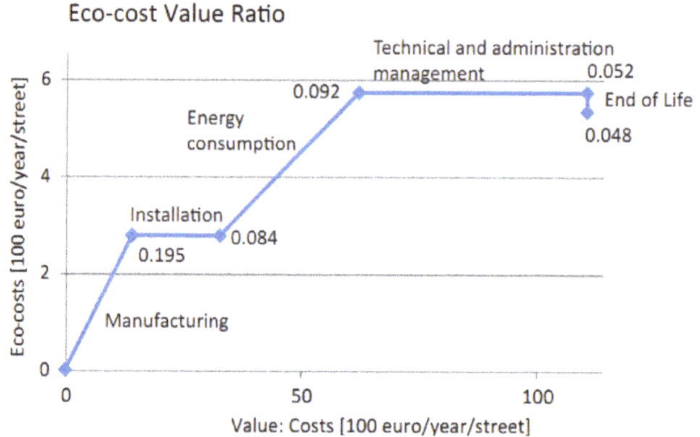

Figure 11. The 2-dimensional representation of costs and eco-costs of the base case (one street, one year). The absolute EVR is provided at each point of the curve.

From this graph it can be concluded where improvements should be made. The 'manufacturing' and 'energy consumption' phase cause the biggest rise in eco-costs, so it makes sense to focus on these issues to reduce eco-costs in the design process. Costs can be saved mainly in 'installation', 'energy consumption', and 'technical and administration management'.

3.2. The Design of the New System

At the idea generation phase of the design, several ways were investigated to fulfil the functional requirements. This is the phase where designers look at all kinds of materials (look and feel [48], recycled or bio-based, shapes (Nature-Inspired Design) [47], and systems (C2C, Life Cycle Thinking). Designers focus on maximum value for the stakeholders. User groups are asked for their preferences.

The eco-burden of concepts at the idea generation phase are normally dealt with by gut feeling, however, this gut feeling is often not fully in line with the reality of LCA. Since the eco-costs of materials weigh heavy in the total eco-costs of the manufacturing of physical products, the LCA-based materials selection app [43] has been developed to give guidance to the designer. When transport and/or energy in the use phase is important, the LightLCA version of the app is required. With the aid of such an app, the environmental aspects of the design are readily available "at your finger tip", so that the designer can focus on the most important aspect of the design at this stage: the creation of value.

In the case of the street lighting system in Rotterdam, the Customer Perceived Value relative to the base case was tested in a small user group for five design concepts: (1) surrounding light attached to the walls of the houses; (2) bamboo posts; (3) Arc light hanging above the street; (4) lamps attached to trees; (5) rooftop-mounted lamps. The rooftop-mounted lamps, see Figure 12, scored the best. A comparison with the base case is shown in Figure 13.

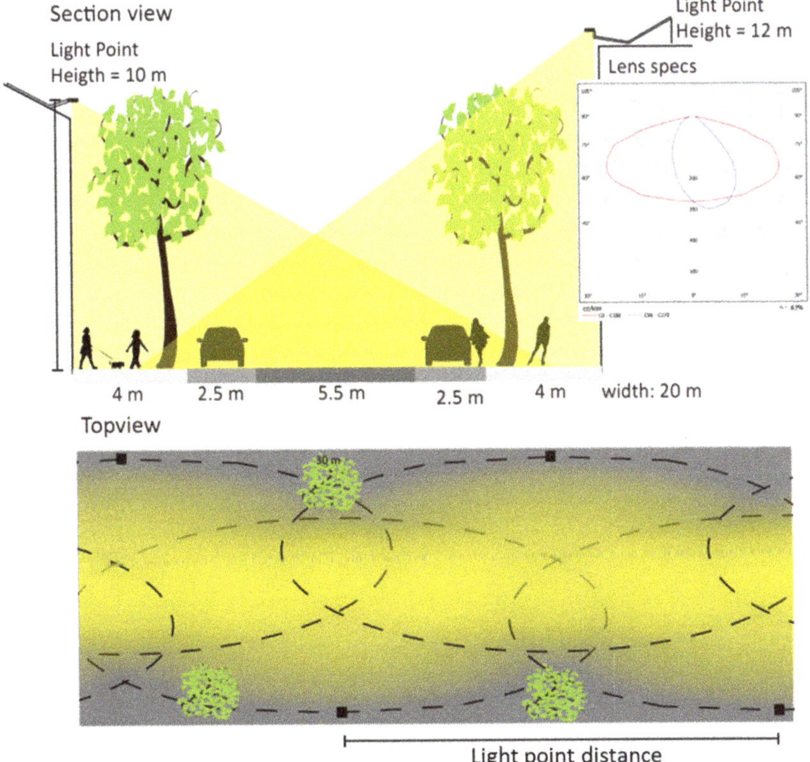

Figure 12. Light design proposal with small and asymmetric beam to avoid light shining into houses.

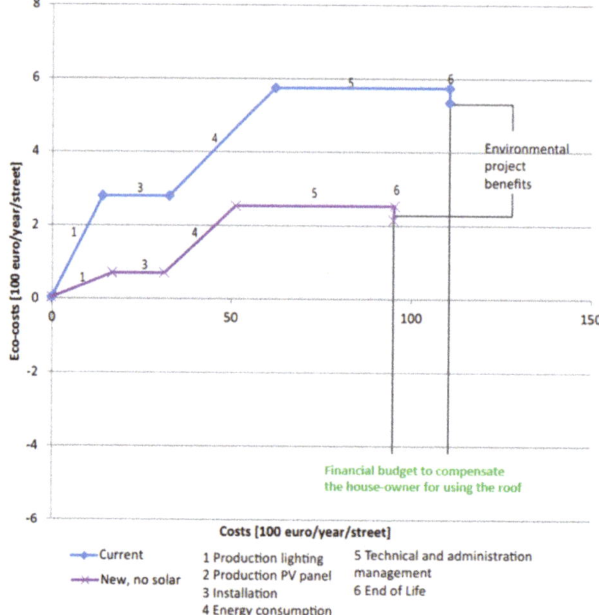

Figure 13. The rooftop lighting system (without PV cell) compared to the base case (one street one year).

An important design issue of Figure 12 is the equal distribution of light, which is a major aspect of the perceived value of street lighting. Shades of shadow cause feelings of unsafety. The combination with trees in the street requires special attention in the system design.

Interesting observations in Figure 13 are: (1) the production costs of the rooftop system are not lower that the production costs of the lamp post system, however, the installation costs are lower; (2) the eco-costs of the rooftop system are considerably lower; (3) the replacement of the PL fluorescent lamps by LED results in less electricity (less costs as well as eco-costs); note that these savings could have been realised with a new lamp post system as well; (4) the benefit of the new system compared to the old system might be used to compensate the house-owner (in this case a housing association) for using the roof.

At a later moment in this project, in the concept development stage, the rooftop-mounted lamps were combined with one PV cell on the roof, a logical system extension in regard to sustainability. The comparison of such a system with the bases case is shown in Figure 14.

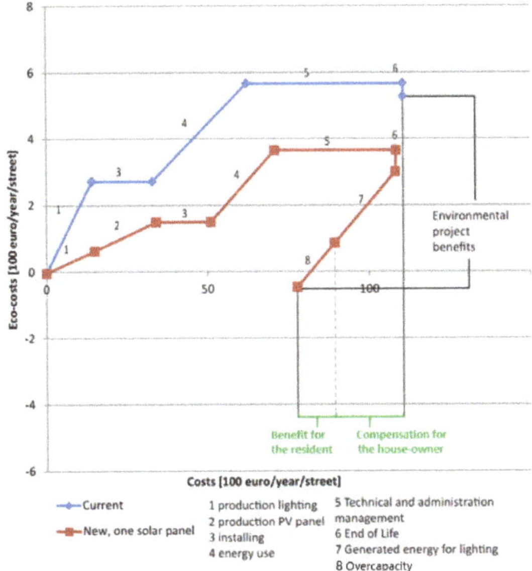

Figure 14. The rooftop lighting system, with one PV cell, compared to the base case.

In the EVR approach, the cost savings of the PV cell (the delivered electricity) is depicted in an extra line at the end of the curve: line 7 + 8. This line has the same slope as line 4, since they are both electricity. At the end of line 7, the amount of electricity that is used by the lamp, is delivered by the PV cell. Line 8 depicts the overproduction of the PV cell. An interesting issue of Figure 15 is how to divide the benefit of a lower Total Costs of Ownership of the new system (compared to the base case) between the house-owner and the resident of the building (to compensate for the extra burden caused by the municipality). Such a division is arbitrary, and will result from negotiations, but the point between line 7 and line 8 might be a logical choice: the benefit for the resident is the overproduction of the PV cell.

Figure 15. A prototype of set 2 PV cells plus lamp.

It is obvious that the owner of the building, in this case a housing association, might take the opportunity of installing extra PV cells. That is kept outside this analysis, but is shown in Figure 15: the first prototype of a set of 2 PV cells plus lamp. This prototype has been redesigned for a test pilot in the Marconistraat 43 in Rotterdam (an industrial area). The test pilot is still operational.

4. Discussion and Conclusions

This project of street lighting systems reveals two important issues:

- The sustainable innovation is not necessarily found in the application of radically new technologies, products or services. Sustainable innovation is the way in which existing technologies, products and services constitute a new sustainable product-service system with a viable business model that adds value to all the three stakeholders: (1) the municipality, by more value for the same costs; (2) the citizens in the street, by adding safety at night in combination with the trees in the street; (3) the owner and/or residents of the building, by reducing the costs of electricity.
- the chosen solution actually has potentially a spin-off effect that might become even more important than the system itself: the design concept *inspires* end-users to place additional, privately owned, solar panels on their roofs, alongside the solar panel of the municipality. Note that this is a very cost-effective way, since the installation of extra panels hardly adds to the installation costs.

Fossil Energy-saving systems (e.g., insulation, heat pumps, windmills, PV cells), have the characteristic that the TCO is less than the investment costs. In the EVR charts, this is characterised by a line with a negative slope, since there are savings in eco-costs as well as costs (see Figure 14). These savings are developing over time. After the pay-back period of the system, the extra cost savings will have a rebound effect [49], since these savings will result in other expenditures (e.g., on cars or holidays). When the EVR of such an expenditure is more than the EVR of the savings, the net result is negative for the environment. When the EVR of the expenditure is less than the EVR of the savings, the net result for the environment is positive [49]. The net result of energy savings has, therefore, a behaviour aspect.

Products for consumer markets must have a surplus value at the moment of purchase, whereas in cases of non-profit organisations, the non-profit organisation has an intermediate position between the stakeholders that pay for the project, and the stakeholders that benefit from the project. As a consequence, eco-efficient value creation for non-profit organisations like a municipality, has two distinct project phases: (1) the choice of the system concept, and the trade-offs between the value (in this case the CPV of the citizens), the eco-burden, and the costs (TCO), are done prior to the start of the implementation project, leading to a budgetary TCO limit. (2) after the project approval, this TCO limit will restrict the further design freedom, however, the approach of eco-efficient value creation still continues for designing further details: creating maximum value at minimum eco-costs. The same situation exists for big infrastructural projects and building design.

The design of a new concept of domestic street lighting system for the city of Rotterdam is a practical example of the approach of Eco-efficient Value Creation. The new concept results in a considerable reduction of carbon footprint and eco-costs, shows the benefits for the municipality and for the residents, and results in a viable business case. The end-result might seem logical and obvious, as it is the case for many good innovations. For all parties that were involved in the design, however, it was clear that such an achievement was the result of the well-structured design process in combination with the establishment of the CPV and eco-costs for several design alternatives in the early design stages (see Supplementary Materials). The design project won the Future Ideas Thesis Competition.

Supplementary Materials: https://www.ecocostsvalue.com/EVR/img/references%20ecocosts/Nine%20Klaassen%20Report.pdf and https://www.ecocostsvalue.com/EVR/img/references%20ecocosts/Nine%20Klaassen%20Appendices.pdf.

Author Contributions: Conceptualisation, N.K., B.F., A.S.; methodology, B.F., A.S.; writing—original draft preparation, N.K.; writing—review and editing, J.V.; All authors have read and agreed to the published version of the manuscript.

Funding: This research received no external funding.

Conflicts of Interest: The authors declare no conflict of interest.

References

1. McDonough and Braungart. *Cradle to Cradle: Remaking the Way We Make Things*; North Point Press: New York, NY, USA, 2002.
2. Ellen MacArthur Foundation. Towards the Circular Economy. Available online: www.ellenmacarthurfoundation.org (accessed on 30 August 2014).
3. McKinsey. The Circular Economy: Moving from Theory to practice Special Edition, October 2016. Available online: https://www.mckinsey.com/~{}/media/McKinsey/Business%20Functions/Sustainability/Our%20Insights/The%20circular%20economy%20Moving%20from%20theory%20to%20practice/The%20circular%20economy%20Moving%20from%20theory%20to%20practice.ashx (accessed on 15 April 2020).
4. Linder, M.; Williander, M. Circular Business Model Innovation: Inherent Uncertainties. *Bus. Strategy Environ.* **2017**, *26*, 182–196. [CrossRef]
5. Geissdoerfer, M.; Vladimirova, D.; Evans, S. Sustainable business model innovation: A review. *J. Clean. Prod.* **2018**, *198*, 401–416. [CrossRef]
6. Oghazi, P.; Mostaghel, R. Circular Business Model Challenges and Lessons Learned—An Industrial Perspective. *Sustainability* **2018**, *10*, 739. [CrossRef]
7. Vogtlander, J.G.; Scheepens, A.E.; Bocken, N.M.P.; Peck, D. Combined analyses of costs, market value and eco-costs in circular business models: Eco-efficient value creation in remanufacturing. *J. Rem.* **2017**, *7*, 1–17. [CrossRef]
8. Scheepens, A.E.; Vogtlander, J.G.; Brezet, J.C. Two life cycle assessment (LCA) based methods to analyse and design complex (regional) circular economy systems. Case: Making water tourism more sustainable. *J. Clean. Prod.* **2016**, *114*, 257–268. [CrossRef]
9. White, A.L.; Stoughton, M.; Feng, L. *Report. Servicing: The Quiet Transition to Extended Product Responsibility*; U.S. Environmental Protection Agency, Office of Solid Waste: Washington, DC, USA, 1999; pp. 1–89.
10. Tukker, A. Eight types of product–service system: Eight ways to sustainability? Experiences from SusProNet. Wiley Online Library. *Bus. Strategy Environ.* **2004**, *13*, 246–260. [CrossRef]
11. Shluzas, L.A.; Steinert, M.; Katila, R. User-Centered Innovation for the Design and Development of Complex Products and Systems. In *Design Thinking Research: Building Innovation Eco-Systems*; Stanford University, Springer Science & Business Media: Berlin/Heidelberg, Germany, 2014; pp. 135–149. ISBN 978-3-319-01303-9.
12. Brezet, J.C.; Van Hemel, C. *Book Ecodesign: A Promising Approach to Sustainable Production and Consumption*; United Nations Environment Programme, Industry and Environment, Cleaner Production: Paris, France, 1997.
13. Mestre, A.; Vogtlander, J.G. Eco-efficient Value Creation of Cork products: An LCA based method for design intervention. *J. Clean. Prod.* **2013**, *57*, 101–114. [CrossRef]
14. Welsh, B.C.; Farrington, D.P. Effects of Improved Street Lighting on Crime. *Camb. Systematic. Rev.* **2008**. [CrossRef]
15. Haans, A.; de Kort, Y.A.W. Light distribution in dynamic street lighting: Two experimental studies on its effects on perceived safety, prospect, concealment, and escape. *J. Environ. Psychol.* **2012**, *32*, 342–352. [CrossRef]
16. Kardan, O.; Gozdyra, P.; Misic, B.; Moola, F.; Palmer, L.J.; Paus, T.; Berman, M.G. Neighborhood Greenspace and Health in A Large Urban Center. *Sci. Rep.* **2015**, *5*, nr 11610. Available online: https://www.nature.com/articles/srep11610 (accessed on 15 April 2020). [CrossRef]
17. Schusler, T.M.; Weiss, L.; Treering, D.; Balderama, E. Research note: Examining the association between tree canopy, parks and crime in Chicago. *Landsc. Urb. Plan.* **2017**, *170*, 309–313. [CrossRef]
18. Huijbregts, M.A.J.; Steinmann, Z.J.N.; Elshout, P.M.F.; Stam, G.; Verones, F.; Vieira, M.D.M.; Zijp, M.; Hollander, A.; van Zelm, R.E.F. ReCiPe2016: A harmonised life cycle impact assessment method at midpoint and endpoint level. *Int. J. Life Cycle Assess.* **2017**, *22*, 138–147. [CrossRef]
19. Lehmann, A.; Bach, V.; Finkbeiner, M. Product environmental footprint in policy and market decisions: Applicability and impact assessment. *Integr. Environ. Assess. Manag.* **2015**, *11*, 417–424. [CrossRef] [PubMed]
20. Environmental Priority Strategies (EPS). Available online: https://www.ivl.se/english/startpage/pages/our-focus-areas/environmental-engineering-and-sustainable-production/lca/eps.html (accessed on 15 April 2020).

21. Vogtlander, J.G.; Baetens, B.; Bijma, A.; Brandjes, E.; Lindeijer, E.; Segers, M.; Witte, J.P.M.; Brezet, J.C.; Hendriks, C.F. *LCA-based Assessment of Sustainability: The Eco-costs/Value Ratio (EVR)*; Delft Academic Press: Delft, The Netherlands, 2010; ISBN 9789065622334.

22. Sala, S.; Cerutti, A.K.; Pant, R. *Report. Development of a Weighting Approach for the Environmental Footprint*; Publications Office of the European Union: Brussels, Belgium, 2019; Available online: https://ec.europa.eu/jrc/en/publication/development-weighting-approach-environmental-footprint (accessed on 30 June 2020).

23. Olindo, R.; Schmitt, N.; Vogtlander, J.G. Electricity in LCA: Inconvenient uncertainties in leading databases, and the need for extra calculation rules. Cases: Calculations on battery electrical cars and hydrogen from electrolysis. *Sustainability* **2009**. Unpublished.

24. Vogtlander, J.G.; Bijma, A. The virtual pollution prevention costs '99: A single LCA-based indicator for emissions. *Int. J. Life Cycle Assess.* **2000**, *5*, 113–124. [CrossRef]

25. Vogtlander, J.G.; Brezet, J.C.; Hendriks, C.F. The Virtual Eco-costs '99, a single LCA-based indicator for sustainability and the Eco-costs/Value Ratio (EVR) model for economic allocation. *Int. J. LCA* **2001**, *6*, 157–166. [CrossRef]

26. Vogtlander, J.G.; Bijma, A.; Brezet, J.C. Communicating the eco-efficiency of products and services by means of the Eco-costs/Value Model. *J. Clean. Prod.* **2002**, *10*, 57–67. [CrossRef]

27. Vogtländer, J.; Peck, D.; Kurowicka, D. The Eco-Costs of Material Scarcity, a Resource Indicator for LCA, Derived from a Statistical Analysis on Excessive Price Peaks. *Sustainability* **2019**, *11*, 2446. [CrossRef]

28. ISO 14008:2019. *Monetary Valuation of Environmental Impacts and Related Environmental Aspects*; International Organization for Standardization: Geneva, Switzerland, 2019.

29. The Model of the Eco-costs/Value Ratio (EVR): The concept of the eco-costs. Available online: https://www.ecocostsvalue.com/EVR/model/theory/subject/2-eco-costs.html (accessed on 15 April 2020).

30. ISO 14044:2006. *Environmental Management—Life Cycle Assessment—Requirements and Guidelines*; International Organization for Standardization: Geneva, Switzerland, 2006.

31. The Model of the Eco-costs/Value Ratio (EVR): Data on eco-costs, Carbon Footprint, and Other indicators: The Idematapp and Idemat Databases. Available online: https://www.ecocostsvalue.com/EVR/model/theory/subject/5-data.html (accessed on 15 April 2020).

32. Porter, M.E. *Competitive Advantage: Creating and Sustaining Superior Performance*; Simon and Schuster: New York, NY, USA, 1985; ISBN 9781416595847.

33. Vogtlander, J.G.; Mestre, A.; Van de Helm, R.; Scheepens, A.; Wever, R. *Eco-efficiënt Value Creation, Sustainable Strategies for the Circular Economy*, 2nd ed.; Delft Academic Press: Delft, The Netherlands, 2014; ISBN 9789065623683.

34. Gadiesh, O.; Gilbert, J.L. Profit Pools: A Fresh Look at Strategy. *Harv. Bus. Rev.* **1998**, *76*, 139–148.

35. McDougall, G.H.G.; Levesque, T. Customer satisfaction with services: Putting perceived value into the equation. *J. Serv. Market.* **2000**, *14*, 392–410. [CrossRef]

36. Lin, C.; Sher, P.J.; Shih, H. Past progress and future directions in conceptualizing customer perceived value. *Int. J. Serv. Ind. Manag.* **2005**, *16*, 318–336. [CrossRef]

37. Gautama, N.; Singhb, N. Lean product development: Maximizing the customer perceived value through design change (redesign). *Int. J. Prod. Econ.* **2008**, *114*, 313–332. [CrossRef]

38. EU Joint Research Centre. Environmental Impact of Products: EIPRO, Analysis of the Life Cycle Environmental Impacts Related to the Final Consumption of the EU-25. Joint Research Centre, European Commission. 2006. Available online: https://ec.europa.eu/environment/ipp/pdf/eipro_report.pdf (accessed on 15 April 2020).

39. Huppes, G.; Ishikawa, M. A Framework for Quantified Eco-efficiency Analysis. *J. Ind. Ecol.* **2005**, *9*, 25–41. [CrossRef]

40. Caiado, R.G.G.; de Freitas Dias, R.; Mattos, L.V.; Quelhas, O.L.G.; Leal Filho, W. Towards sustainable development through the perspective of eco-efficiency—A systematic literature review. *J. Clean. Prod.* **2017**, *165*, 890–904. [CrossRef]

41. Huguet Ferran, P.; Heijungs, R.; Vogtländer, J.G. Critical Analysis of Methods for Integrating Economic and Environmental Indicators. *Ecol. Econ.* **2018**, *146*, 549–559. [CrossRef]

42. Lofthouse, V. Ecodesign tools for designers: Defifining the requirements. *J. Clean. Prod.* **2006**, *14*, 1386–1395. [CrossRef]

43. Available online: www.idematapp.com (accessed on 15 April 2020).

44. Vogtlander, J.G. A practical Guide to, L.C.A. In *For Students, Designers and Business Managers; Cradle-to-crave and Cradle-to-cradle*, 5th ed.; Delft Academic Press: Delft, The Netherlands, 2017; ISBN 9789065623614.

45. Rivas-Hermann, R.; Köhler, J.A.; Scheepens, A.E. Innovation in product and services in the shipping retrofit industry: A case study of ballast water treatment systems. *J. Clean. Prod.* **2015**, *106*, 443–454. [CrossRef]

46. Osterwalder, A.; Pigneur, Y.; Bernarda, G.; Smith, A. *Value Proposition Design*; Wiley: London, UK, 2014; ISBN 9781118968055.

47. Van Boeijen, A.; Daalhuizen, J.; Zijlstra, J.; Van de Schoor, R. *Delft Design Guide*; BIS Publishers: Amsterdam, The Netherlands, 2017; ISBN 9789063693275.

48. Karana, E. Meanings of Materials. Ph.D. Thesis, Delft University of Technology, Delft, Netherland, 29 May 2009. Available online: https://repository.tudelft.nl/islandora/object/uuid:092da92d-437c-47b7-a2f1-b49c93cf2b1e (accessed on 15 April 2020).

49. Scheepens, A.E.; Vogtlander, J.G. Insulation or Smart Temperature Control for Domestic Heating: A Combined Analysis of the Costs, the Eco-Costs, the Customer Perceived Value, and the Rebound Effect of Energy Saving. *Sustainability* **2018**, *10*, 3231. [CrossRef]

© 2020 by the authors. Licensee MDPI, Basel, Switzerland. This article is an open access article distributed under the terms and conditions of the Creative Commons Attribution (CC BY) license (http://creativecommons.org/licenses/by/4.0/).

Review

Reviewing ISO Compliant Multifunctionality Practices in Environmental Life Cycle Modeling

Christian Moretti [1], Blanca Corona [1], Robert Edwards [2], Martin Junginger [1], Alberto Moro [2], Matteo Rocco [3,*] and Li Shen [1]

[1] Copernicus Institute of Sustainable Development, Utrecht University, 3584 CB Utrecht, The Netherlands; c.moretti@uu.nl (C.M.); b.c.coronabellostas@uu.nl (B.C.); h.m.Junginger@uu.nl (M.J.); l.shen@uu.nl (L.S.)
[2] Joint Research Centre, European Commission, 21027 Ispra, Italy; robert.edwards@ec.europa.eu (R.E.); alberto.moro@ec.europa.eu (A.M.)
[3] Department of Energy, Politecnico di Milano, 21056 Milan, Italy
* Correspondence: matteovincenzo.rocco@polimi.it

Received: 27 May 2020; Accepted: 6 July 2020; Published: 11 July 2020

Abstract: The standard ISO 14044:2006 defines the hierarchical steps to follow when solving multifunctionality issues in life cycle assessment (LCA). However, the practical implementation of such a hierarchy has been debated for twenty-five years leading to different implementation practices from LCA practitioners. The first part of this study discussed the main steps where the ISO hierarchy has been implemented differently and explored current multifunctionality practices in peer-reviewed studies. A text-mining process was applied to quantitatively assess such practices in the 532 multifunctional case studies found in the literature. In the second part of the study, citation network analysis (CNA) was used to identify the major publications that influenced the development of the multifunctionality-debate in LCA, i.e., the key-route main path. The identified publications were then reviewed to detect the origins of the different practices and their underlying theories. Based on these insights, this study provided some "food for thought" on current practices to move towards consistent methodology. We believe that such an advancement is urgently needed for better positioning LCA as a tool for sustainability decision-making. In particular, consistent allocation practices could be especially beneficial in bioeconomy sectors, where production processes are usually multifunctional, and where current allocation practices are not harmonized yet.

Keywords: bibliometrics; review; life cycle assessment (LCA); allocation; system expansion

---

1. Introduction

Life cycle assessment (LCA) is supposed to be a standardized methodology to measure the life cycle impacts of products or services. LCA is currently ruled by ISO 14040:2006 and ISO 14044:2006 [1,2]; these standards have been the basis of the LCA methodology for the last two decades. Nevertheless, in the scientific community, some experts wonder if the detail presented in these standards is enough to guide LCA practitioners in practice [3–5].

One of the most debated problems in LCA is the so-called "multifunctionality" issue (or commonly, "allocation") [6–8]. Multifunctionality issues need to be dealt with when different product systems share a process, e.g., manufacturing processes delivering more than the studied product, or end-of-life activities providing both waste management service and a recovered or recycled product. In these cases, apportioning environmental burdens among the co-products, or rather co-functions, becomes necessary. According to ISO 14044:2006, multifunctionality should be solved by using the following three-level hierarchy [2]:

1. Avoiding allocation by subdivision (dividing the unit process into two or more sub-processes) or system expansion ("expanding the product system to include the additional functions related to the co-products");
2. Allocation following underlying physical relationships (i.e., an allocation that quantitatively reflects how the inputs and outputs are changed by changes in the amount of each product of the system);
3. Allocation (partitioning) based on other relationships (e.g., economic value).

The same hierarchy applies also to "open-loop" recycling, i.e., when a material is recycled as a different product because it is no longer suitable to replace the original product directly. Only in such open-loop recycling, ISO 14044:2006 provides further guidance on the third level of the hierarchy, where physical properties (e.g., mass) are preferred to economic value, which in turn is preferred to the number of subsequent uses of the recycled material [2].

The existence of the ISO's multifunctionality hierarchy should avoid the use of inadequate approaches, e.g., determined by the interests of the stakeholders or the ones of the study's commissioner [3]. Nevertheless, the apparent lack of sufficient guidance has fed different implementation practices [9]. Consequently, although most LCAs claim compliance with the two ISO standards, practitioners have applied different allocation procedures in LCAs assessing the same or similar products [10]. Since the choice of the allocation method typically affects the outcome of the LCA significantly [6,10–13], this problem has led to different conclusions and therefore low reliability and robustness of the LCA results [14]. Moreover, due to the lack of a shared view in the LCA community, some authors decide not to follow the ISO hierarchy (see [15]), while other authors select the allocation method based on their subjective decision (see, e.g., [16] and [17]). Other researchers choose allocation methods that are "commonly" applied in similar case studies in the literature (see, e.g., [18]), others calculate also an average allocation parameter considering common parameters (e.g., [19]) or others use "conservative" allocation methods that provide the highest impacts (e.g., see [20]).

This article presents a literature review on the main practices and debates on using ISO 14044:2006 recommendations to solve multifunctionality problems. A critical literature review on multifunctionality methodology development was combined with quantitative analysis of current multifunctionality practices, and a bibliometric review based on citation network analysis (CNA). The quantitative analysis was performed by a text-mining process in 532 multifunctional case studies found in the literature.

The CNA was used to identify the main knowledge flow on multifunctionality in LCA, also known as "the main path". Tools and software based on the "main path" method are used for many applications: tracking the evolutionary trajectory of a science field or the development of a specific technology, or the evolving changes of legal opinions of courts [21,22]. The "main path" was investigated to detect the historical origins of the different practices currently present in the literature and their underlying theories. The use of such a tool overcomes some limitations of the traditional systematic reviews conducted so far on this topic, which were based on "human" selection of the articles (e.g., through criteria such as the number of citations).

In the literature, the definitions used to characterize the multifunctionality issue are not harmonized. For this reason, we provided Appendix A reporting the definitions used in this review to distinguish the different types of products, multifunctional processes, modeling approaches and system expansion approaches.

2. Methodology

Figure 1 summarizes the three main steps followed in this literature review. First, the literature search was performed. Second, a critical review was conducted to identify the main issues and bottlenecks in the LCA literature when implementing the ISO allocation procedures. The critical review was combined with a text-mining process to quantitatively assess the current practices in the LCA literature (focusing on all the LCA case studies selected by the query). Third, a bibliometric analysis was performed based on citation network analysis (CNA).

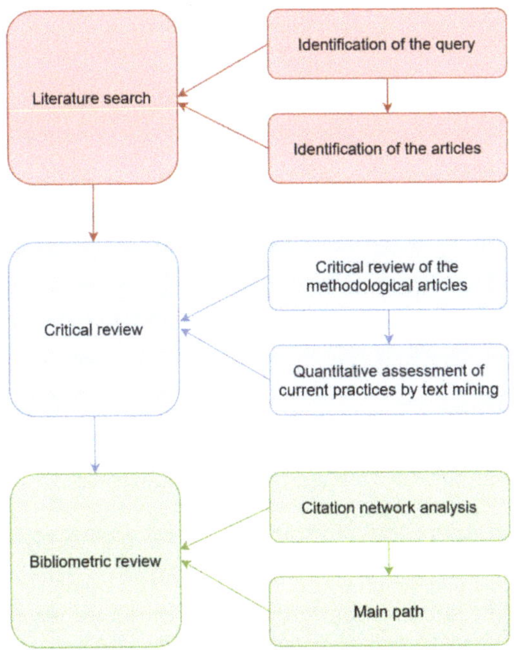

Figure 1. The three steps followed in the literature review.

The literature search was based on data collected from the Scopus database in February 2019. The searched publication fields were: title, abstract, and keywords. The search string was characterized by the terms: "Life Cycle Assessment", "LCA", "multifunctionality", "allocation" and "multi-output". Since allocation approaches are also used in other fields (e.g., in business management), the query was first limited to environmental assessment or engineering-related fields. Because of this, the documents were reduced from 1310 documents to 1152. This allowed us to exclude 145 documents belonging to business management, 6 related to veterinary science and 7 others. Our analysis was further refined by considering articles only from the category of Scopus "journals". By applying this last adjustment, the articles resulting from the search became 930. Since only research articles were analyzed, some relevant books or conference proceedings may have been excluded from the analysis. Nevertheless, books often resume the contributions previously published as articles, and some excluded documents might have been considered by some of the reviewed reviews. Figure 2 shows the number of publications per year, highlighting the growing interest in the topic.

The corpus of documents on which the analyses were performed included the 930 articles retrieved from Scopus and the main LCA guides and standards, i.e., ISO technical reports and standards (also withdrawn ones like ISO 14041:1998) [23–26], the International Reference Life Cycle Data System (ILCD) handbook [27], the Product Environmental Footprint (PEF) guide [28] and the Product Environmental Footprint Category Rules (PEFCR) guidance [29]. Out of the 930 documents, 307 studies were identified through their title and abstract as "methodological articles" (of which 117 were review articles focusing on a specific sector where LCA is applied). These methodological articles focused either on the general methodological debate about multifunctionality procedures, or discussed a specific method, or introduced a new model to solve multifunctionality. The most relevant articles in this group were critically reviewed to understand the main issues when solving multifunctionality in LCA while claiming compliance with ISO. This critical review focused mainly on the articles cited more than 20 times ("most cited ones") and the articles published after 2015 ("recent ones").

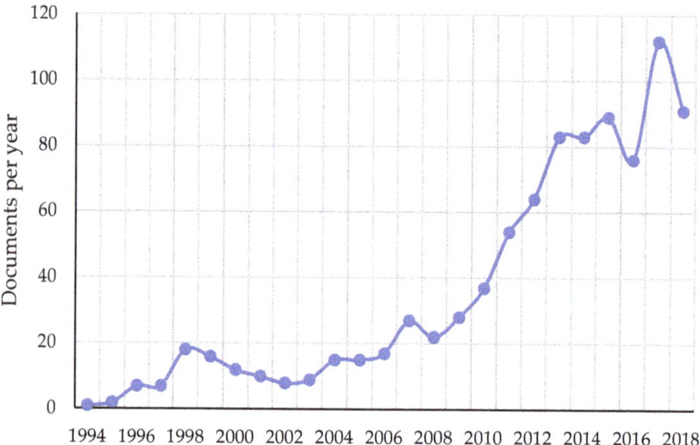

Figure 2. The time distribution of the articles on multifunctionality in LCA published in scientific journals per year retrieved from Scopus.

The critical review was combined with a text-mining process whose aim was to quantify the current practices when solving multifunctionality issues. The text-mining process was manually performed on the remaining 532 case studies. These 532 case studies resulted from a further refinement which excluded 91 articles that either did not apply full LCA or were not environmental LCA studies. Concerning the multifunctional case studies retrieved from the literature, we observed that specific parts of the bioeconomy namely agriculture (63 case studies), bioenergy (185), bio-based materials (52) and anaerobic digestion (21), were the ones most affected by the issue of multifunctionality together with related sectors namely aquaculture (14), dairy and meat products (79), fossil counterparts (34) and waste management (50). These sectors together represented 94% of the 532 case studies identified by the query.

Text-mining software can detect relevant terms or keywords in the corpus of literature with less time and cost than a person [30]. However, when the keywords represent technical concepts, dedicated software typically achieves low to medium efficiencies (e.g., 25%–65%) [30]. For instance, software could not understand when the concept "system expansion" was used as an alternative expression for substitution or for system enlargement. To increase the efficiency of the text-mining method, the quantitative estimation was performed directly by the analyst. When the terms representing the concepts of interest (e.g., "allocation") were encountered, the context of their use was assessed by reading the surrounding text.

In the third step, i.e., the bibliometric review, the 930 articles were investigated by CNA. The CNA was performed using Pajek software [31]. Documents are considered "nodes" and the citations are the "links" between these documents. The type of nodes is defined therefore based on the type of document. The "sources" are the documents that are cited but cite no other documents and therefore represent the origins of the knowledge. The "sinks" are the documents that cite other documents but are not cited and therefore could represent the "current stage" of the knowledge stream. Intermediate documents cite other previous documents and are also cited by more recent documents [21]. Our CNA aimed at identifying the main path of research. This path represents the main knowledge flow in a specific topic, i.e., the major contributions that have influenced the development of the research, which does not mean directly the most cited ones overall [32,33]. The main path was obtained by using an algorithm that computed what citations between articles had been more significant. In particular, such a significance was calculated through the key-route method [34]. This method identifies the main chain of articles by considering the highest transversal count [33,34]. The transversal counts measure the significance of a citation link, i.e., by counting the times a citation link is traversed [34]. The transversal count adopted was the search path count (SPC). The SPC assigns as value to each link the number of paths traversing the link among all possible paths connecting all the sources to all the sinks [21,22].

3. The Critical Review Combined with Text Mining

When critically reviewing the methodological articles on multifunctionality, it emerged that these articles present two main "debates" regarding ISO-compliance practices. These two debates concern the application of system expansion (explained in Section 3.1 and related sub-Sections) and the identification of relevant partitioning criteria (see Section 3.2). In particular, Pelletier et al. (2015) identified three "schools" distinguished by the way they interpret the ISO hierarchy with respect to these three aspects: (1) the consequential LCA (CLCA) thinking school interprets system expansion as substitution, (2) the natural-science attributional school applies system expansion as enlargement and prioritizes allocation based on a physical parameter, and (3) the socio-economic attributional school applies system expansion as enlargement but prefers economic allocation. According to Pelletier et al. [9], these three schools are "internally consistent" but "mutually exclusive".

3.1. Debate on the Interpretation of ISO's System Expansion

The system expansion debate focuses on how and when the substitution method should be applied. ISO 14044:2006 recommends system expansion as a way to avoid allocation, but no further specification is provided regarding the differences between enlargement and substitution (see Appendix A for detailed definitions), and about its implementation in attributional or consequential LCAs. Substitution is often used as a system expansion approach in attributional LCAs (ALCAs), which is not perceived as correct by many LCA experts [5,9,35–38]. According to these practitioners, ALCA modeling should not rely on perturbation logic or counterfactual notions, such as substitution or avoidance of other products/processes (as also highlighted, e.g., by Majeau-Bettez et al. [35]). It is argued that the sum of the impacts accounted by attributional LCAs should add up to the worldwide impacts, and this would not be valid anymore if substitution were applied [5,39]. For this reason, Chen et al. (2010) concluded that the "allocation methods, even if perfectible, are still preferable to the system expansion method" (used as synonymous of substitution), because "system expansion does not ensure a global coherency between various LCA studies" [40]. On this basis, the use of substitution as a system expansion method in ALCA is not supported by any of the schools of interpretation identified by Pelletier et al. [9]. Similarly, Bailis and Kavlak, after applying substitution for the by-products of a biofuel, concluded that "the large disparity between system expansion and other methods raises questions about the validity of system expansion" [41]. Concerning system expansion by enlargement, this cannot be applied when the goal of the study requires the impacts of just one of the co-products or by-products to be obtained. In these cases, allocation cannot be avoided. For example, "In a milk production system that also produces beef, system expansion without substituting would lead to a system with a function of delivering both milk and beef" [42].

Other authors argue that ISO 14044:2006 does not acknowledge substitution as a system expansion approach. The reason is that ISO refers only to the addition of functions (i.e., enlargement) and not to the substitution of functions [3,5,43–45]. On these bases, several authors argue that a distinction of ALCA/CLCA should be present in future ISO 14044 [9,46] since, for them, this distinction is crucial to select the appropriate system expansion method (enlargement or substitution) (as also pointed out by [47,48]). By contrast, other authors argue the opposite, i.e., that substitution is generally recognized as a valid method for avoiding allocation within attributional LCA [49,50]. For many practitioners, substitution is considered as synonymous with system expansion [51,52]. Under this argument and considering the ISO hierarchy, substitution should be preferred to any allocation method [53–57]. Pelletier et al. [9] suggested that the equivalence substitution-system expansion might have originated in a 1994 study authored by Tillman et al. [58]. The reason was that Tillman et al. [58] is a frequent citation when justifying the equivalence of substitution with system expansion. However, their study was published prior to the publication of the ISO standards.

3.1.1. Current Practices in Specifying the Modeling Approach

Although the choice of modeling approach (ALCA or CLCA), which depends on the goal, clearly determines the outcome of an LCA study, our text-mining process found that the keywords "attributional" and "consequential" were missing in 75% of the LCAs involving multifunctional systems (see Figure 3 for a detailed breakdown, per product sector). This percentage refers only to the portion of articles published after 2004 when the term consequential LCA was clearly established (see Section 4.3). There are several possible reasons for this low specification rate of the modeling approach: (1) practitioners could still not be aware of the relevance to differentiate between consequential and attributional approaches, (2) practitioners may not specify the modeling approach because it is a direct consequence of the goal description, (3) they may not agree with a strict distinction between ALCA and CLCA, (4) they may be strictly following current ISO standards that do not distinguish between the two approaches or (5) they may have followed the recommendations of a policy directive or national/international guide that does not make such a distinction. Actually, some ALCA studies combined with consequential thinking are emerging [10]. These approaches aim mainly at accounting for some specific counterfactual effects or credits, and at the same time, limit complexity and uncertainties [59].

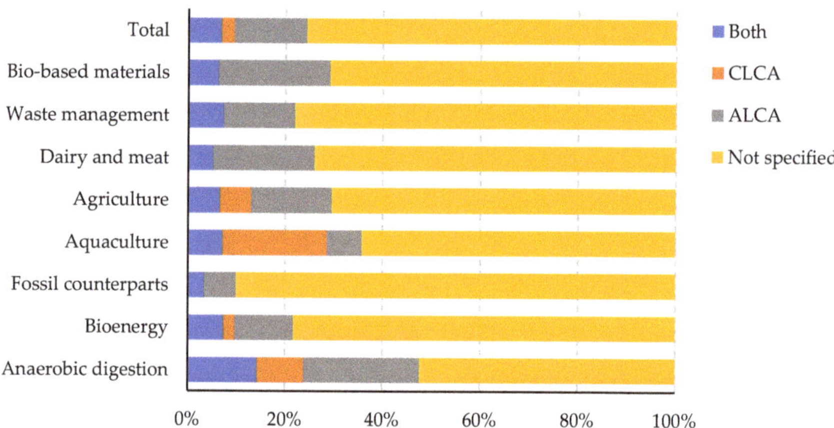

Figure 3. Percentage of articles which applied on the same case study both CLCA and ALCA approaches (Both), self-declared attributional studies (ALCA), self-declared consequential studies (CLCA) and studies which did not declare the approach followed (Not specified). Only case studies published after 2004 were considered (504).

The text-mining process found that 31% of the self-declared ALCA studies (using the keyword "attributional") used substitution as a system expansion approach to avoid allocation. However, this percentage varies depending on the sector under consideration, ranging from 19% to 45% (see Figure 4). The highest rate of substitution approaches in ALCAs was found in studies related to bio-based materials (45%). On the other hand, there are few LCAs investigating fossil products that self-declared as attributional studies. The reason might be that substitution is rarely an option for fossil products since they are usually the "substituted products".

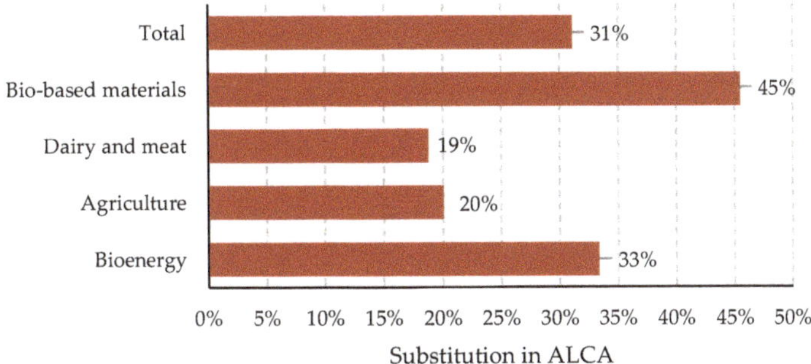

Figure 4. Percentages of self-declared attributional studies (ALCAs) which applied substitution as a system expansion approach. Only the sectors with a significant amount (more than 10) of self-declared attributional studies are included in this graph.

3.1.2. The Application of Substitution as System Expansion Method

Beyond the use of substitution in ALCA, two other critical aspects of substitution have been discussed in the LCA methodological articles. First, the high uncertainties introduced by the use of the substitution approach since it can lead to different results depending on the choice of substituted and/or substituting by-product [3,60,61]. A sensitivity analysis should be therefore recommended. Second, when substitution is suitable, the substitution of co-products should be avoided by checking the physical/economical significance of the products delivered by the multifunctional process [3,27]. However, some authors (for example [3]) argue that the importance of the co-products' physical significance is not emphasized enough in ISO 14044:2006 and the ILCD handbook. When physical significance is not checked and a by-product is credited for the replacement of co-products, the practitioner could obtain significantly distorted results [3,62]. A common practice to account for physical significance is to select the primary functions based on the main source of revenues [35,63]. In cases where the primary co-function(s) cannot be directly identified, the ILCD handbook proposes that they should be assumed to be those that jointly contribute to more than 50% of the combined market value of all co-functions of the analyzed multifunctional process [27].

Clear rules for differentiating by-products from the co-products are important, because, in substitution, all the credits from substituting by-products are attributed to the main co-product. If another LCA on the same process is made in which a by-product is considered to be the main product, the impacts of the process "get counted twice", so that the impacts for different products no longer add up to the total for the process (this would be a problem in an attributional model—for further details see the next Section).

3.1.3. Using Substitution as the Allocation Method

Another point of debate is the use of substitution for allocation (and not as a system expansion method). This type of allocation has been mentioned in the literature with different names, such as substitution-based allocation [9] and "proxy-based disaggregation" by substitution [35] and various versions of this method have been proposed (e.g., see Hermansson et al. who applied two different versions of this method to assess Kraft lignin [64]). By many practitioners, this option is perceived as the attributional way of using substitution. PEFCR guidance and PEF guide [28,29] propose that, when a by-product of a multifunctional system directly substitutes another product, such substitution might be considered as an allocation reflecting physical relationships. When this is the case, such substitution has to be based on a direct and empirically demonstrable relationship [28,65]. Pelletier et al. [65] stated that this is different from substitution based on marginal market models applied in consequential LCAs [65]. An example of such a substitution is when "manure nitrogen is applied to agricultural land, directly substituting an equivalent amount of the specific fertilizer nitrogen that the farmer would otherwise have applied" [28,65]. Hence, it is assumed that the impact caused in the system by the production of the substituted by-product corresponds to the impact of the production of the replaced product (as shown in [66]). With substitution, the impact of a by-product should equal that of the product it substitutes, and so is independent on the actual process that produces it. Moreover, the application of this substitution-based allocation can lead to a negative impact that in ALCA would mean that the model has been built inconsistently [35]. As an example, this happens when the wrong product is chosen as the main product of the multifunctional system [66] or the substituted product is not a minor product, even if representing less than 50% of market value [67]. Even if the substituted co-products are chosen carefully (i.e., they represent small percentages in physical and economical terms), this method sometimes fails in ALCAs assessing multiple impact categories, resulting in negative impacts for some of them [13,67]. Moreover, PEFCR guidance and PEF guide [28,29] also allow the possibility of using indirect substitution as a form of allocation based on "other relationship". "Indirect substitution may be modeled as a form of allocation based on some other relationship when a co-product is assumed to displace a marginal or average market-equivalent product via market-mediated processes" [28].

3.2. Selection of the ISO Allocation Criterion

The main discussion on the allocation criterion concerns the nature of the so-called ISO "physical relationships" and "other relationships" [7,42].

The authors in line with the socio-economic school argue that allocation can be based on physical relationships only when the ratio of the output products can be varied since this allows the establishment of physical causality between functional units by mathematical modeling [68–73]. For example, Bernier et al. [74] assessed the impact of Kraft lignin and applied the physical causality principle to allocate the impact between pulp and lignin "by varying the quantity of lignin precipitated and then observing direct variations in the environmental loads". They also specified that this type of allocation was selected based on ISO standards, which recommends this type of allocation over allocation based on mass, energy or economic values. This school, therefore, interprets "physical relationships" as "physical causality relationships" and interpret "other relationships (e.g., economic value)" as "other causal relationships". Accordingly, they consider the allocation by other relationships as the only possible approach when it is not possible to change the ratio of production of the functional outputs of the system [69,72]. The practitioners following the view of this school often argue that, at this level, economic allocation is the recommended option, and only when it is not possible to use economic allocation, the allocation can be based on a physical parameter that should be selected based on the best proxy for economic revenues (e.g., see [7]). For example, this happens when there is a lack of market prices for one specific product [75]. However, the approximations of these causal relationships have always been a debated scientific issue [76]. For example, these relationships can be based on the common function of all co-products (as done by [77]).

The text-mining process revealed that only 28% of the LCA case studies selected an allocation method based on ISO relationships interpreted as "causal relationships". The percentage of studies following this interpretation varied significantly depending on the sectors considered (see Figure 5). In particular, it was very low in the studies focusing on anaerobic digestion, bioenergy and bio-based materials (5%–16%). On the other hand, this interpretation is largely present in the fossil fuels sector, where the allocation of emissions to single products is often based on linear programming models calculating marginal emissions by varying the amount of functional units [78,79]. Another example where this interpretation is largely present is in the dairy sector. The main reason is that many practitioners assessing dairy products often selected their allocation choice based on the recommendations of the International Dairy Federation [80], which adopts this interpretation.

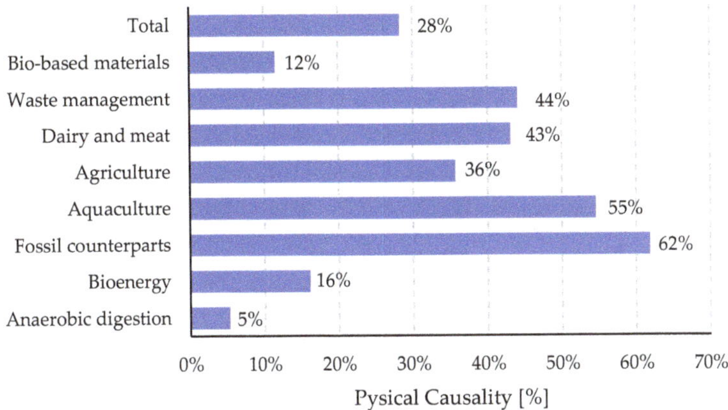

Figure 5. The percentage of studies in each research area that used causality as the principle for the allocation choices per (sub-)cluster and overall. The number of studies per sector: anaerobic digestion (21), bioenergy (185), fossil counterparts (34), agriculture (63), aquaculture (14), dairy and meat (79), waste management (50), bio-based materials (52).

Conversely, the practitioners belonging to the natural-science school often refer to an allocation by physical parameter as ISO-second level allocation by interpreting "physical relationships" as allocation based on a physical parameter, e.g., mass or energy value [8,16,81–83]. On this basis, they prefer allocation based on a physical parameter (e.g., mass) over economic allocation because "ISO 14044 standard mentions economic allocation when no other possibility is available" [84]. The economic allocation may be selected by the practitioners following this view if allocation based on physical parameter "result in the attribution of a large proportion of burdens to low-value co-products" [9]. The same school often argued that an allocation based on a physical parameter is preferred over economic allocation since it is not affected by price fluctuations [82,85]. As a response, authors in line with the socio-economic school argue that the price fluctuation is not the important parameter for the allocation method, but the ratio of prices among all products, which is much less variable because it mainly depends on the fluctuating price of the common inputs to the process [86].

The preference expressed by the natural-science school is adopted by PEFCR guidance and PEF guide, which prefer allocation based on physical keys (e.g., mass or energy) to economic ones [28,29]. In the PEF guide, ISO "physical relationships" might have been interpreted as allocation based on physical parameters (this emerges from our understanding of annex X of PEF guide), leading to the preference for physical allocation keys. On the contrary, the ILCD handbook adopts the interpretation of "ISO physical relationships" from the socio-economic school and states that only when it is not possible to find clear physical *causal* relationships between the co-functions, allocation based on economic relationships can be used [27]. However, differently from what is usually preferred by the socio-economic school, the ILCD handbook does not give preference to economic allocation over non-causal physical properties such as energy content [27]. The ILCD handbook also adds a footnote to remark that energy allocation is not an allocation based on ISO causal physical relationship but a simplified allocation based on a physical property that is not causal [27].

To make an example of the implication of adopting one interpretation or the other, we can consider a biorefinery example that produces fuels (e.g., ethanol) and chemicals for materials (e.g., lactic acid) [87]. The natural-science school would prefer energy or mass allocation (considered by them as ISO second level) over economic allocation (considered ISO third level). Conversely, the socio-economic school would prefer economic allocation, arguing that mass and energy allocations (all considered ISO third level) are meaningless for such a biorefinery because of not representing any causality mechanism. They would also argue that it is not appropriate to use energy allocation when not all the co-products are used for their energy content or to use mass allocation when there are energy products among the co-products.

4. The Bibliometric Review Based on Main Path Analysis

The main path of research identified using CNA is shown in Figure 6 and included 21 articles. The evolution of the multifunctionality discussion in the scientific community can be divided into four periods, which have been defined as (1) bilateral beginning, (2) the ISO 14041 influence, (3) consequential LCA influence, and (4) ISO 14044 application.

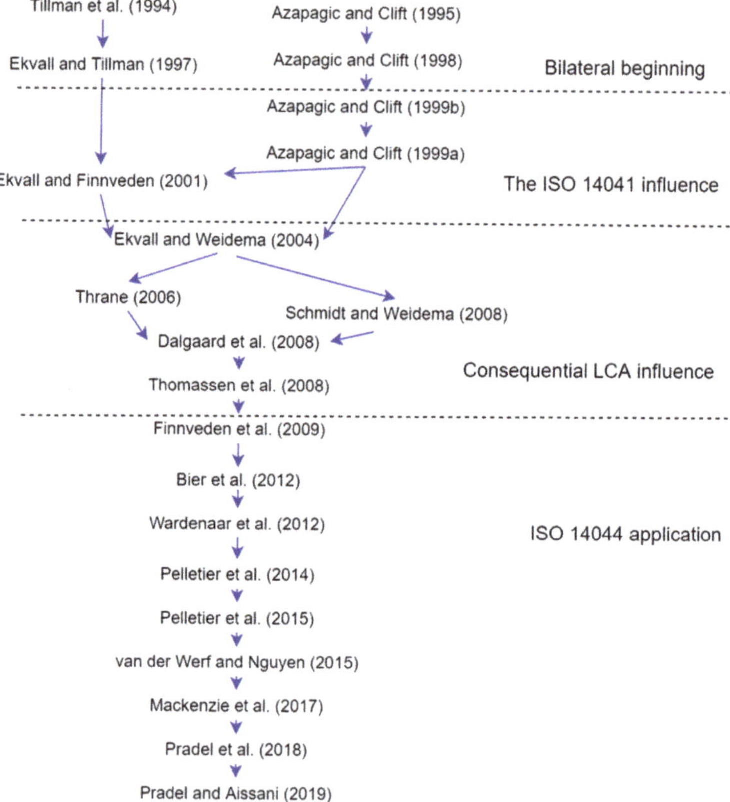

Figure 6. The key-route main path of research on LCA multifunctionality (output from Pajek calculations), obtained from the citation network analysis.

4.1. Bilateral Beginning (1994–1998)

The discussions on the LCA multifunctionality issue were initially developed following two parallel routes (see Figure 6). On the first route, Tillman, Ekvall and their co-authors developed different types of allocation methods for multi-output systems and open-loop recycling [58,88]. It is crucial to notice that at that time, the ISO 14041 was not yet released [25]. Tillman et al. [58] focused their article on the choice of system boundaries based on the purpose of an LCA. They defined three LCA purposes: 1) process tree (PT), today known as ALCA and applied to processes where there are one main product and some by-products, 2) technological whole system (TWS), similar to what today is known as ALCA and applied to processes delivering several co-products, and 3) socio-economic whole system (SWS), similar to current CLCA [58]. In this article, the word *expansion* was used once with respect to SWS, indicating that such a system accounts for economic and social factors and therefore "may lead to further expansion of the system" [58]. In 1994 and 1996, two conferences were held with sessions on allocation and life cycle inventory. Clift, who was also co-author of the publications in the second parallel route, published the reports of such sessions [89,90]. These reports concluded that allocation must, when possible, be based on causal relationships. Ekvall and Tillman discussed this conclusion, arguing that causal relationships could be either cause-oriented or effect-oriented [88]. An example of the first one is the manufacture of a product that occurs because the company expects customers to be willing to pay for it (cause). An example of an effect-oriented relationship is a system delivering a recycled product, which reduces the amount of virgin product in another system (effect). This second type of relationship resembles the current CLCA thinking. To represent effect-oriented relationships, they argue that the effects of the investigated product on other life cycles can be included in the LCA through the expansion of system boundaries. As the expansion of the system boundaries, they cited the approach developed by Tillman et al. (1994), which today is known as "substitution". Moreover, they argued that when LCA is used as a tool for decision support, the allocation procedure should generally be effect-oriented rather than cause-oriented. Therefore, it is possible to identify the probable origin of the consequential school in the study of 1997 of Ekvall and Tillman [88] and the study of 1994 of Tillman et al. [58].

The four articles of the second route were authored by Azapagic and Clift. In the first article (1995), they proposed linear programming (LP) modeling to solve the multifunctionality issue and to calculate the optimized environmental impact of plastic resins production, such as polypropylene and polystyrene [91]. The inputs and outputs of the system are then allocated to each of the co- and by-products through marginal changes in its production [91]. The marginal allocation coefficients correspond to the variation of the environmental burdens associated with a marginal variation of one of the co-functions [91]. The second article (1998) focused as well on LP as a tool for solving the problem of allocation and was applied to systems producing borate products. They highlighted that 1) "the main characteristic of this kind of modeling is that it is based on physical and technical relationships between the inputs and outputs […] describing the underlying physical causation in the system" [73] and 2) that the allocation by causal relationships provided by the model is obtained "by exploring how the burdens change when the quantity of one function is changed with the quantities of all the other functions kept constant" [73]. These changes can be marginal, incremental, or average ones; however, LP can only be applied when system behavior can be linearized, which does not usually happen in average changes (i.e., substantial changes as for example the elimination of a functional output completely) [73].

In 1996, the first draft [92] of the ISO hierarchy for solving multifunctionality was released, as reported by Ekvall and Tillman [88]. This hierarchy was very similar to the one still present in the current ISO 14044:2006. System expansion was indicated in the first level [92]: "by expanding the system boundaries so that inputs, outputs and recycles remain within the system" (retrieved from [88]). From the literal statement, it appears clear that it was intended as an enlargement of the system boundaries to include all the co-functions within the boundaries (see Figure A1 in Appendix A). Such an approach is different from the system expansion method (substitution) indicated by Tillman's SWS, where functions are avoided instead of added.

On the second level, it was stated [92]: "where allocation cannot be avoided, the allocation should be based on the way in which the inputs and outputs are changed by quantitative changes in the products or functions delivered by the system" (retrieved from [88]). There was no use of the term "physical relationships" as in ISO 14044:2006. Hence, ISO was proposing allocation methods such as the marginal allocation developed by Azapagic and Clift [91], which are based on quantitative changes in the products or functions delivered by the system.

Finally, the last level allowed the allocation of different functions based on economic relationships, excluding allocation by physical properties. This preference for economic values could be due to their cause-oriented essence (the function is provided because one is willing to pay for it). Based on the analysis of this bilateral beginning and this ISO draft [92], it seems that the socio-economic ALCA school represents the first version of the ISO hierarchy. In fact, they distinguish themselves by applying system expansion by only adding (and not subtracting) functions, and preferring economic allocation to an allocation based on a physical parameter (excluded option by this first draft version).

4.2. The ISO 14041 Influence (1999–2003)

In the third article of the second route, Azapagic and Clift used the boron product system to examine the different allocation methods recommended by ISO 14041, which was just released in 1998 [25,93]. ISO 14041:1998 introduced the same three levels of the hierarchy of the present ISO 14044:2006. However, the second level included the following clarification: "the resulting allocation will not necessarily be in proportion to any simple measurement such as mass or molar flows of coproducts". Azapagic and Clift argued that, following ISO 14041, allocation by physical (causal) relationships had to be the result of mathematical system modeling [70,93]. Nevertheless, the ISO 14041 allocation underlying physical relationships allowed also allocations based on the "cause of the limits" of the amount of product output. This aspect emerges from the annex of ISO 14041:1998, where mass or volume allocations are suggested as representing physical relationships for road transportation because the quantity of materials transported is limited by the maximum load that the vehicle can carry [25]. Although these two approaches may at first appear contradictory, they are in line with Azapagic and Clift's work, who also concluded that in some cases (which include the transportation example), allocating based on a physical quantity leads to the same results obtained by marginal allocation [70,93]. In these cases, it may be correct to allocate based on a physical parameter representing the physical causation involved, and therefore, not arbitrarily [70,93].

Azapagic and Clift (1999a) highlighted that system expansion (enlargement) is not always applicable. This approach is not possible when the goal of the study requires to determine the impacts of only one of the products [93]. The reason is that, by expanding the functional unit to include the co-functions, the results at the level of one single product would not be available. They also investigated allocation in heat and power cogeneration plants. Due to lack of data, they could not model the system to represent physical causalities and therefore applied the "avoided burdens approach" (in later research "substitution"). Azapagic and Clift argued that substitution is a conceptually equivalent alternative to system expansion, and is suitable when one co-product displaces its production elsewhere, such as for energy recovery from waste or cogeneration [93].

Actually, annex B of ISO 14041 quoted the same example of system-expansion/substitution applied to energy from waste incineration [25]. Nevertheless, annex B specified that the expansion of the boundaries like this requires 1) that the goal of the study is aimed at assessing a change, "i.e., a comparison between two alternative scenarios for the same product" and 2) that the modeled change which will actually occur because of the decision supported by the LCA can be predicted with a fair degree of certainty [25]. To apply this type of expansion, the LCA should aim, therefore, to answer the question of what would have been the long-term marginal effect if the service had not been performed [25]. Hence, substitution became a possible system expansion approach in what nowadays is understood as consequential thinking. This annex with allocation examples is no more included in the current ISO 14044:2006.

In the lowest level of the allocation hierarchy of ISO 14041 [25], economic allocation became an example, and no more the only acknowledged allocation method as it was reported in the previous draft version [92]. Hence, in some cases, allocation based on a physical parameter could be preferred to an economic allocation, and this might have given birth to the natural-science ALCA school. Moreover, ISO 14041:1998 specified that the environmental impact should be allocated only to the products causing the release of the emissions (causality principle). ISO 14041:1998 proposed the example of a multi-input incineration process releasing cadmium emissions which should be allocated only to the input wastes that contain cadmium.

The fourth article [70] referred once again to LP-based marginal allocation, stating that this modeling applies "when the functional outputs can be varied independently", i.e., in partial joint production or combined production (see Appendix A for more details about these definitions). A naphtha cracking was proposed as an example of a system where the outputs can be independently varied (within physical and thermodynamic limits) by changing the operating conditions [70]. When that is not the case (i.e., full joint production; with a fixed ratio of products), "allocation by physical causality cannot be implemented" [70]. Linked to this impossibility, they provided the often-cited example of the ratio of sodium hydroxide (NaOH) and chlorine (Cl2) produced by electrolyzing brine, which is fixed by stoichiometry. Other examples that they mentioned about this impossibility are rapeseed oil/residue (ratio fixed by the chemical structure of the plant) and beef/leather (fixed by the physical structure of the animal) [70]. In these cases, the authors stated that ISO recommended economic allocation because it reflects "the socio-economic demands which cause the multiple-function systems to exist" [70]. They concluded that "allocation on an arbitrary basis, such as mass or energy flow, must be avoided" and "where physical causality between functional units and environmental burdens exists, the allocation should always be based on these causal relationships" [70]. The authors based their methodological choices on the 1997 voting draft of ISO 14041.

The ISO 14041:1998 was complemented by ISO/TR 14049:2000 [23]. This technical report defined system expansion as the addition of functions but lost the concept of system expansion with substitution when the goal is to assess a change. This is still missing in the current ISO 14044:2006 and ISO/TR 14049:2012 [2,24].

This ISO/TR provided two examples related to the disposal phase of the life cycle. The first example showed how to expand boundaries to compare two processes with different outputs, A and B, using the same inputs. As illustrated in case d of Figure A1 in Appendix A, the system boundary for each process needs to be expanded with an alternative process for making the other product. Then the two systems under comparison produce the same functional unit A+B. Moreover, it specified that the added processes shall be those that "would actually be involved when switching between the two analyzed systems" [23]. In the second example, open-loop recycling is solved with a closed-loop procedure that includes the entire recycling processes into the same system boundaries (like case c of Figure A1 in Appendix A).

Concerning allocation by physical property (e.g., mass or viscosity), ISO/TR 14049:2000 [23] emphasized that this type of allocation should be preferred to economic allocation only when it reflects the way in which the inputs and outputs are changed by quantitative changes in the products, (as, for example, in the transportation example in ISO 14041:1998, quoted above). This had to be proven by varying the ratio of co-products [23].

In 2001, Ekvall and Finnveden published a critical review on allocation in ISO 14041:1998 [72]. Ekvall and Finnveden stated that system expansion (in the form of substitution) could be used in a broader range of LCA goals than the one for which it is recommended by the annex of ISO 14041. For example, it can be used to account for indirect effects [72], similar to how the substitution method is used today in CLCAs.

In the same review, Ekvall and Finnveden (2001) identified the marginal allocation of [70] as a method corresponding to the second level in the ISO hierarchy (the first connection between the two parallel routes in Figure 6). In particular, Ekvall and Finnveden [72] explained that there were two possible interpretations for ISO allocation based on physical–causal relationships. Under the first interpretation, the "environmental burdens allocated to a function should be the burdens avoided if that function is no more delivered while the other functions are unaffected" [72]. This type of allocation is applicable when the environmental burdens are linear with the quantity of each of the functions delivered and, therefore, it is possible to eliminate the functions independently [72]. The second interpretation is that "the environmental burdens allocated to each of the functions should be proportional to the partial derivatives at the point of operation" [72]. This is a generalized description of the LP modeling of Azapagic and Clift.

Concerning the third level of ISO hierarchy, they emphasize that a rigorous interpretation of the standard leads to an allocation based on other causal relationships, e.g., economic value, and not in non-causal relationships (e.g., allocation based on "arbitrary physical property of the products such as mass, volume or energy content") [72]. As aforementioned, this strict approach was also the only one foreseen in the first draft of ISO hierarchy [92] and favored by the ALCA socio-economic school. At this point, the main path stopped to be bilateral and started a period of interconnection that led to the development and definition of what today is categorized as "consequential thinking".

4.3. Consequential LCA Influence (2004–2008)

In 2004, the keyword *consequential LCA* appeared for the first time on the main path [94]. In the same article, Ekvall and Weidema delineated the consequential LCA as commonly defined today. They stated that CLCA avoids allocation by applying substitution-type system expansion, using marginal data [94].

Following the main path of research, we found several articles on CLCA case studies. In the first article, Thrane conducted a CLCA of fish products [95]. The second article authored by Schmidt and Weidema [96] is focused on how to identify the marginal vegetable oil to be substituted in a CLCA of agricultural systems providing food and oil. Thrane [95] pointed out that, generally, the ISO allocation hierarchy can also be considered valid for CLCA. In fact, when system expansion (either by enlargement or by substitution) or subdivision is not applicable, it is also necessary to allocate by physical or other relationships in CLCA [95]. Dalgaard et al. [97] then performed a CLCA of soybean meal and avoided allocation by applying the substitution of marginal vegetable oil [97]. The fifth article of this period, authored by Thomassen et al., compares attributional and consequential LCAs of milk production [98]. They showed that depending on the modeling approach (ALCA or CLCA), the results significantly vary for the same system because of the different ways of dealing with multifunctionality (allocation versus system expansion with substitution). In the middle of this period, ISO 14044:2006 was released.

4.4. ISO 14044:2006 Application (from 2009)

At this stage, the most cited article identified by the search was published by Finnveden et al. [99]. This article repeated that the underlying physical relationships of ISO14044:2006's second hierarchy level should represent physical, chemical, or biological causation (as also specified before in [73]). Consequently, economic, mass, or energy allocations were intended to be used only as the third level option [99].

Following the main path, we found an LCA on a bio-based plastic product derived from a blood meal [100]. Bier et al. highlighted how different approaches for solving allocation issues in LCAs of bio-based materials could widely vary the results. The next two articles of the main research path discussed the choice of allocation approaches to use in LCAs aimed at informing policy-making. Wardenaar et al. [101] pointed out that methodological uncertainty within ISO led to significantly different results due to the influence of the allocation approach, and argued that the policy context could benefit from new guidelines [101]. Concerning the ISO hierarchy, they stated that "several authors have argued that substitution is equivalent to system expansion" referring as an example to [88]. However, "conceptually equivalent does not mean that system expansion and substitution provide the same results" because there are "large differences between these two methods" [101]. As a consequence of this assumed conceptual equivalency, some authors "use this implicit argument to choose for substitution, while still claiming compliance to ISO" [101]. Concerning allocation based on a physical parameter, Wardenaar et al. argued that the physical parameter should be the one reflecting the physical characteristics related to "the purpose or use of the product", i.e., the relevant characteristic for which they are sold [101].

Following the main path, we found a study on the Environmental Footprint guidelines published by the European Commission [65]. The study of 2014 of Pelletier et al. [65] highlighted that in ISO's first level system expansion, the functional unit is expanded to include the other co-functions (enlargement), and the impacts are therefore reported at the system level, i.e., at the level of all co-products [65]. This was claimed to be the "literal interpretation of ISO 14044" [9]. Accordingly, the PEF guide [28] does not consider substitution as a system expansion approach, but only enlargement (similarly the more recent PEFCR guidance [29]). However, the ILCD handbook allows system reduction as an option in CLCA, and for those ALCAs whose aim is to include also the interactions with other systems [27].

The key-route main path analysis allowed us to identify the origins of the "equivalency" substitution-system expansion, i.e., the articles on the side of the bilateral beginning period originated by Tillman et al. [58]. The suggestion of Pelletier et al. [9] that this equivalency originated from the 1994 study of Tillman et al. [58] was therefore confirmed by our analysis. Nevertheless, the article by Tillman et al. [58] was published before any ISO standard and, therefore, did not refer to the system expansion method as intended by ISO.

In the next publication of the main path, Pelletier et al. (2015) observed that, despite the ISO hierarchy, consistent implementation of this hierarchy in the literature was limited, and presented the three schools of thought (consequential, socio-economic ALCA and natural-science ALCA) mentioned at the beginning of Section 3.

The next two articles in the main research path were focused on finding allocation parameters for agricultural systems. The first article proposed an allocation based on plant physiological construction cost for plant compounds, which should represent the underlying physical relationships between co-products, i.e., the physiological mechanism involved in plant growth [102]. Hence, they concluded that, according to ISO, such a method should be preferred to allocation based on common properties of co-products, such as energy or economic content [102]. Subsequently, Mackenzie et al. [69] studied similar biophysical allocation methods and concluded instead that these methods might not represent the causal physical mechanisms of these systems because they overlook the interconnectivity between co-products [69] as instead, an LP would do. Therefore, they concluded that allocation by economics is preferable [69]. Mackenzie et al. also pointed out that many practitioners often choose an allocation based on an arbitrary parameter (e.g., their mass or energy content) also when it does not reflect such a cause-effect mechanism [69].

The last two articles of the main path were focused on how to allocate burdens to by-products which were previously considered wastes [103,104]. These by-products are scarce wastes that can be converted into valuable products. In particular, Pradel et al. constructed a novel allocation method based on relevant causal relationships obtained by mathematical modeling [103]. This model was applied to wastewater treatment plants delivering sludge (by-product) and clean water (main product) and calculated the allocation factor for sludge and water.

5. Discussion and Conclusions

Despite the existence of a hierarchy for solving multifunctionality in ISO 14044:2006, the complexity of the multifunctionality problem, the lack of sufficient guidance, its difficult interpretation, and the discrepancies in other "ISO-compliant" guides or handbooks have led to a wide variety of allocation procedures in the literature. Such variety is especially present in the system expansion approaches and in the choice of the allocation key.

ISO 14044:2006 does not distinguish between attributional and consequential modeling. For many practitioners, distinguishing between attributional or consequential LCAs is a crucial key to selecting the method to deal with multifunctionality. For other practitioners, some mixed approaches can be considered as advancements in the methodology. We found that only 25% of the LCAs clearly state the approach followed using the terms "attributional" and "consequential". Are practitioners not specifying it because they assume it to be "intrinsically clear" from the goal description, or because they do not agree with such a distinction? Some mixed approaches have also been proposed in the literature.

The first major reason for debate on ISO's multifunctionality hierarchy is the application of substitution as a system expansion method in ALCA (found in 31% of the self-declared attributional studies explored through text mining). Such practice is perceived as inappropriate by many practitioners.

However, some practitioners who do not acknowledge substitution as system expansion in ALCA recognize the use of substitution as an allocation method for ALCA. Concerning the use of substitution, another aspect that many practitioners pointed out is that a future ISO standard should emphasize more the criterion of physical/economical significance as a prerequisite to apply substitution to avoid incorrect interpretations of the results.

The second reason for the debate is the meaning and application of the "ISO relationships" criterion for the selection of the allocation method. A first interpretation (found in 28% of the case studies) is that the ISO refers to "causal physical relationships" as relationships mathematically modeled, while "other causal relationships" relate to other relationships (e.g., based on physical or economic parameters) selected based on the best proxy for physical relationships. The second interpretation is that allocation by "physical relationships" refers to an allocation by physical parameters (e.g., mass or energy) while "other relationships" refer to economic relationships.

Most (94%) of the LCAs of multifunctional case studies found in the literature search are linked to bioeconomy (agriculture, biofuels, bioenergy, and biomaterials) and its linked sectors (fossil fuels and petrochemical plastic materials and dairy products). This has generated inconsistencies within each area, but also at the boundaries between these sectors, because of their multiple links. As an example, biogas can be produced from the manure of a farm, which produces dairy products with animals that eat dried distillers' grains with solubles coming from ethanol fermentation. Such ethanol production may have a pre-treatment process shared with lactic acid fermentation. This lactic acid may be used to produce poly-lactide which, in the market, replaces polypropylene. The above biogas can then be used to generate electricity that can be partly consumed on the farm and partially injected in the grid, substituting power from fossil fuels. How much double counting or how many inconsistencies arise when ISO 14044:2006 is interpreted differently in each of these sectors?

The bibliometric review based on the analysis of the main path obtained from tracing the citation network allowed us to (1) reconstruct how the implementation practices of the ISO hierarchy developed in the last 25 years, (2) identify the origin of the different interpretations and their rationales, and (3) understand how the discrepancies found in the critical review were generated. It emerged that, originally, the ISO hierarchy [92] recommended the approach followed by the "socio-economic ALCA school". The socio-economic ALCA school interprets system expansion as enlargement but prefers economic allocation to allocation based on physical parameters representing a proxy for causality. The origin of the "natural-science ALCA school" was traced to ISO 14041:1998 [25], when allocation by physical parameter as well as economic allocation was permitted as an example of ISO "allocation by other relationships". The natural-science ALCA school interprets system expansion as enlargement and applies allocation based on a physical parameter (for a part of the practitioners subscribing to this view, this choice is justified only when a physical parameter representing causality principles is identified). Its role was promoted by the release of the PEF guide and PEFCR guidance [28,29], which expressed a preference for allocation based on physical parameters over economic ones. Another important view is the one of the "CLCA school" interpreting system expansion as substitution and selecting the allocation method based on causality principles. The birth and development of the CLCA school were found in the annex of ISO14041:1998 and in the publications of Ekvall and co-authors [72,88,94]. They were the first ones (in the main path) to acknowledge the suitability of the substitution method to avoid allocation and account for counterfactual effects (originally proposed by Tillman et al. [58]) and the assumption of "conceptual equivalency" of substitution with the system expansion method.

Summarizing, following one or the other school of thought, a different method is often preferred for the same system, goal and decision context. Applying these different methods could lead to different conclusions and, sometimes, opposite conclusions.

To increase the consistency and reliability of LCA, we believe that a future revised ISO should:

1. Clearly state if distinguishing between attributional and consequential LCA is a key principle to implement the hierarchy. If yes, then it should differentiate the hierarchy for the two approaches and clarify if the hierarchy allows substitution as a system expansion method in attributional LCAs.

2. Clarify the meaning of allocation by "physical relationships" and "other relationships", providing more examples and details than the ones reported in ISO 14044:2006 and ISO 14049:2012.

Author Contributions: Conceptualization, C.M. and B.C.; methodology, C.M., B.C., A.M. and M.R.; software, C.M.; validation, C.M., B.C., R.E., M.J. and L.S.; formal analysis, C.M., B.C., R.E. and M.R.; investigation, C.M., R.E., A.M.; resources, A.M., M.R. and L.S.; data curation, C.M., B.C., M.J., L.S.; writing—original draft preparation, C.M., B.C., R.E.; writing—review and editing, R.E., M.J., A.M., M.R. and L.S.; visualization, B.C., L.S.; supervision, B.C., M.J., A.M., M.R. and L.S.; All authors have read and agreed to the published version of the manuscript.

Funding: This research received no external funding

Conflicts of Interest: The authors declare no conflict of interest.

Appendix A

Appendix A.1. Type of Products

In this article, three definitions are used for different types of products and services: *co-products, by-products and wastes.*

Co-products are the ones satisfying the main (primary) function that a production system or process is intended to deliver. As highlighted by Majeau-Bettez et al. [35], co-products have also been defined with the term "primary", "determining", and others. Conversely, by-products represent only secondary functions of the system. A by-product is a substance resulting from a production process whose primary function is not the production of that item but either it is inevitably produced or could, in principle, be avoided by the system without altering the main functionality of the process (e.g., a farm with tourist accommodation services).

The primary function of a product system is identified by evaluating the purpose of such a system. For example, for the internal combustion engine of a car, the primary product is the mechanical power needed by the car to carry people (primary function). A secondary function of the same engine can be the production of heating (by-product) to keep a proper temperature in the car. Nevertheless, the distinction between primary and secondary functions can be particularly difficult for some unit processes (e.g., sunflower oil vs meal). When such difficulty is encountered, the primary function should be selected by assessing what function of the multifunctional process generates more revenues for the investigated process [27,35,63], within the temporal scope of the LCA. Nevertheless, there are processes whose aim is the generation of several functions of comparable value. In such a case, there can be multiple primary functions. For example, a biorefinery can produce various chemicals and fuels as primary functions (co-products) and provide district heating as a secondary function (by-product).

The shared environmental impact of a process shall be apportioned between co-products and by-products, but not to wastes [2]. According to ISO 14044:2006, wastes are "substances or objects which the holder intends or is required to dispose of" [2]. There is, however, a fine line between wastes and by-products. For example, manure is nowadays used as feedstock for biogas plants, used cooking oil is used for biodiesel production, and residues of the potato industry are used for animal feed. When these alternative uses make these wastes find a market demand represented by market values, they should be considered, therefore, as by-products. We adopt the distinction waste/by-product provided by the Waste Directive Framework [105]. A "waste" becomes a by-product when the "following conditions are met: (1) further use of the substance or object is certain; (2) the substance or object can be used directly without any further processing other than normal industrial practice; (3) the substance or object is produced as an integral part of a production process; (4) further use is lawful, i.e., the substance or object fulfills all relevant product, environmental and health protection requirements for the specific use and will not lead to overall adverse environmental or human health impacts" [105].

Appendix A.2. Type of Multifunctional Processes

As highlighted by several authors, the terminology reported in the literature for distinguishing the different types of multifunctional processes is not harmonized [5,35]. This article follows the terminology defined by Majeau-Bettez et al. [35], who differentiated between *full-joint production, partial joint production*, and *combined production*. *Full joint production* takes place when the co-products are produced simultaneously, with a fixed ratio of production (e.g., fixed by the stoichiometry of a chemical reaction, or by natural processes such as the proportions between wheat grains and wheat straw). *Partial joint production* occurs when there is an intermediate level of technological linkage between the different co- and by-products (e.g., an oil refinery as a whole or the production of milk and meat or the transportation of two different products) and *combined production* when there is not technological linkage (e.g., a gasoline station also offering shop services). According to this definition, the ratio of production of the co- and by-products could be varied in every case except for the full joint production.

Appendix A.3. Type of Modeling Approaches

The selection of the modeling approach is based on the goal of the study and the decision context. Generally, when the goal of a study is to describe the status of a system, an attributional LCA (ALCA) approach is followed to calculate the environmental impact of providing a specific amount of the functional unit [5]. When the goal is to describe the effect of a change due to a decision, a consequential LCA (CLCA) approach is followed to estimate how this environmental impact would change in response to a *change* in the output of the functional unit (i.e., it is change-oriented) [98]. The current conceptualization of the CLCA approach was first publicly discussed in the 2001 international workshop on electricity data for life cycle inventories [106].

One of the main principles of ALCA is the so-called 100% additivity [35]. This principle means that "results of a separate analysis of all economic activities should add up to the result of an analysis of the total economic activity" [39], so ALCA is suitable for attributing the total impacts to a defined function (product or service), but, for example, it does not indicate to policy makers the impact of policy changes, when these cause an incremental change from the *status quo*.

By contrast, CLCA determines the *change* in impacts due to a *change* in the production of the product or service, or to a change in policy. So it attempts to consider *all* the impacts of the change, also on other sectors that are influenced, for example as a consequence of the use of by-products [35]. CLCA is therefore preferred to ALCA for estimating the impact of policy changes [107]. CLCA usually uses market-driven modeling to forecast what will happen once the product or service of interest is introduced [59]. This means that in CLCA, marginal processes are considered, rather than average ones, including the activities displaced by by-products. This is typically modeled through the so-called substitution approach, whereby CLCA considers only the activities reacting to the change in demand for the functional unit, keeping the total of other services constant. Therefore, the quantification of displaced activities depends on the market characteristics of competing products [14].

Appendix A.4. Type of System Expansion Approach and Substitution

System expansion means the enlargement of the boundaries of the system under investigation to include additional processes and functions. As mentioned above, expansion of the boundaries can be used to avoid allocation. There are two possible approaches to avoid allocation by expanding the boundaries: enlargement (see Figure A1 for different types of enlargement) and substitution (see Figure A2). By considering the subtraction as "a negative addition" [51], substitution is considered by some LCA practitioners as a form of system expansion used to isolate the impact of just one function from a multifunctional process.

One can apply system enlargement by modifying the functional unit to include all co-functions (case a of Figure A1). This approach is not possible when the goal of the study requires to determine the impacts of only one of the products because the results at the level of one single product would not be available. System enlargement is also often used for comparative assessments. In case b of Figure A1, the aim is to compare process P1 (providing functions A and B) with process P2 (providing only function A). One needs to add to P2 another process for producing B in order to allow the comparison for the same outputs. Similarly, in case d of Figure A1, the aim is to compare a process producing A with a process producing B (for example, comparing the impacts of two products which could be made from the same raw material). In this case, one needs to add alternative processes for making both A and B in order to make a meaningful comparison. Even though these processes are not initially multifunctional, system enlargement is applied to allow for a fair comparison. One can also apply system enlargement in open-loop recycling systems. In the example of system enlargement from ISO/TR 14049, open-loop recycling is solved with a closed-loop procedure that includes the entire recycling processes into the same system boundaries (like case c of Figure A1).

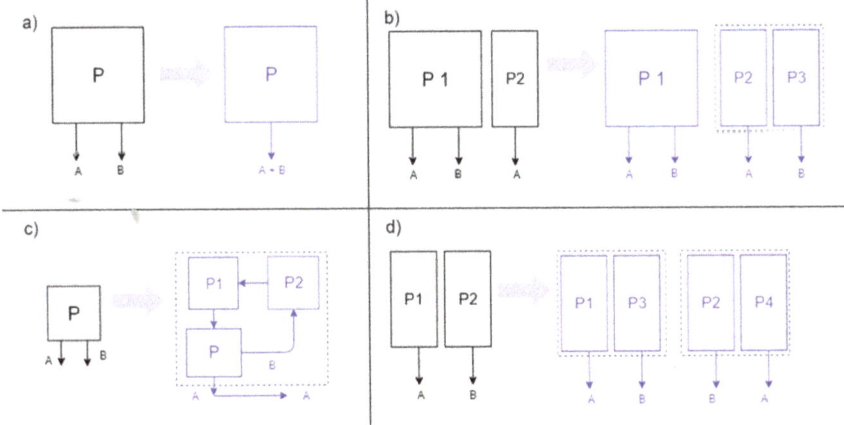

Figure A1. Different ways to apply system expansion as enlargement-addition of functions. In black: multifunctional process before applying system expansion. In blue: process after the expansion of the boundaries/addition of functions. (**a**) Changing the FU to avoid allocation. (**b**) Adding extra processes (P3, delivering B) to a system (P2, delivering A) that is compared with another system (P1) delivering several functions (A and B). (**c**) Applying closed loop recycling to a system (P) where one of the outputs (B) is used as a material input in the same product system (P2 represents the intermediate processing of B that allows its re-use). (**d**) Adding extra processes (P3 and P4) to compare systems that provide different functions and that at the beginning were not multifunctional.

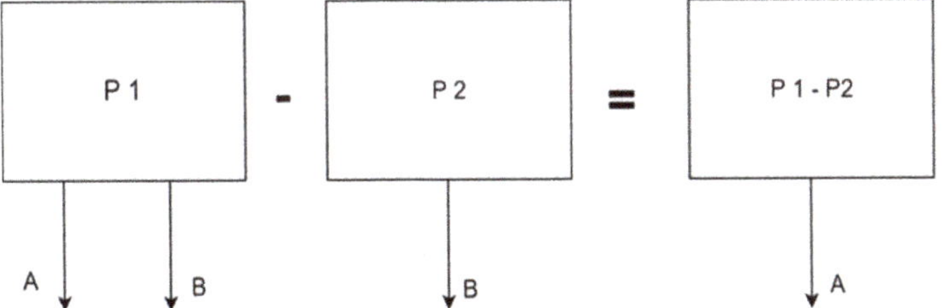

Figure A2. System expansion by substitution (reduction of functions). The investigated system (P1) delivers two products (A and B). Alternatively, product B can be produced by another system (P2). The substitution method proposes that the impact of producing A (only) by process P1, corresponds to the difference of impact between P1 and P2.

References

1. ISO (International Standard Organization). *ISO 14040: Environmental Management—Life Cycle Assessment—Principles and Framework*; Technical Committee ISO/TC 207; ISO: Geneva, Switzerland, 2006.
2. ISO (International Standard Organization). *ISO 14044, Environmental Management—Life Cycle Assessment—Requirements and Guidelines*; ISO/TC 207/SC 5; ISO: Geneva, Switzerland, 2006.
3. Sandin, G.; Røyne, F.; Berlin, J.; Peters, G.M.; Svanström, M. Allocation in LCAs of biorefinery products: Implications for results and decision-making. *J. Clean. Prod.* **2015**, *93*, 213–221. [CrossRef]
4. Zamagni, A.; Buttol, P.; Porta, P.L.; Buonamici, R.; Masoni, P.; Guinée, J.; Heijungs, R.; Ekvall, T.; Bersani, R.; Bieńkowska, A.; et al. *Critical Review of the Current Research Needs and Limitations Related to ISO-LCA Practice*; Enea: Rome, Italy, 2008.
5. Schrijvers, D.L.; Loubet, P.; Sonnemann, G. Developing a systematic framework for consistent allocation in LCA. *Int. J. Life Cycle Assess.* **2016**, *21*, 976–993. [CrossRef]
6. Reap, J.; Roman, F.; Duncan, S.; Bras, B. A survey of unresolved problems in life cycle assessment. Part 1: Goal and scope and inventory analysis. *Int. J. Life Cycle Assess.* **2008**, *13*, 290–300. [CrossRef]
7. Weidema, B. Avoiding Co-Product Allocation in Life-Cycle Assessment. *J. Ind. Ecol.* **2000**, *4*, 11–33. [CrossRef]
8. Muench, S.; Guenther, E. A systematic review of bioenergy life cycle assessments. *Appl. Energy* **2013**, *112*, 257–273. [CrossRef]
9. Pelletier, N.; Ardente, F.; Brandão, M.; De Camillis, C.; Pennington, D. Rationales for and limitations of preferred solutions for multi-functionality problems in LCA: Is increased consistency possible? *Int. J. Life Cycle Assess.* **2015**, *20*, 74–86. [CrossRef]
10. Agostini, A.; Giuntoli, J.; Marelli, L.; Amaducci, S. Flaws in the interpretation phase of bioenergy LCA fuel the debate and mislead policymakers. *Int. J. Life Cycle Assess.* **2019**, *25*, 17–35. [CrossRef]
11. Lloyd, S.M.; Ries, R. Characterizing, propagating, and analyzing uncertainty in life-cycle assessment: A survey of quantitative approaches. *J. Ind. Ecol.* **2007**, *11*, 161–179. [CrossRef]
12. Klöpffer, W. The critical review of life cycle assessment studies according to ISO 14040 and 14044. *Int. J. Life Cycle Assess* **2012**, *17*, 1087–1093. [CrossRef]
13. Moretti, C.; Junginger, M.; Shen, L. Environmental life cycle assessment of polypropylene made from used cooking oil. *Resour. Conserv. Recycl.* **2020**, *157*, 104750. [CrossRef]
14. Brando, M.; Martin, M.; Cowie, A.; Hamelin, L.; Zamagni, A. Consequential Life Cycle Assessment: What, How, and Why? *Encycl. Sustain. Technol.* **2017**, *1*, 277–284.
15. Benetto, E.; Jury, C.; Kneip, G.; Vázquez-Rowe, I.; Huck, V.; Minette, F. Life cycle assessment of heat production from grape marc pellets. *J. Clean. Prod.* **2015**, *87*, 149–158. [CrossRef]
16. Zaimes, G.G.; Khanna, V. The role of allocation and coproducts in environmental evaluation of microalgal biofuels: How important? *Sustain. Energy Technol. Assess.* **2014**, *7*, 247–256. [CrossRef]
17. Śliwińska, A.; Burchart-Korol, D.; Smoliński, A. Environmental life cycle assessment of methanol and electricity co-production system based on coal gasification technology. *Sci. Total Environ.* **2017**, *574*, 1571–1579. [CrossRef]
18. Bava, L.; Bacenetti, J.; Gislon, G.; Pellegrino, L.; D'Incecco, P.; Sandrucci, A.; Tamburini, A.; Fiala, M.; Zucali, M. Impact assessment of traditional food manufacturing: The case of Grana Padano cheese. *Sci. Total Environ.* **2018**, *626*, 1200–1209. [CrossRef]
19. Esteves, V.P.; Esteves, E.M.; Bungenstab, D.J.; Feijó, G.L.; Araújo, O.D.; Morgado, C.D. Assessment of greenhouse gases (GHG) emissions from the tallow biodiesel production chain including land use change (LUC). *J. Clean. Prod.* **2017**, *151*, 578–591. [CrossRef]
20. Pa, A.; Craven, J.S.; Bi, X.T.; Melin, S.; Sokhansanj, S. Environmental footprints of British Columbia wood pellets from a simplified life cycle analysis. *Int. J. Life Cycle Assess.* **2012**, *17*, 220–231. [CrossRef]
21. Liu, J.S.; Chen, H.H.; Ho, M.H.C.; Li, Y.C. Citations with different levels of relevancy: Tracing the main paths of legal opinions. *J. Assoc. Inf. Sci. Technol.* **2014**, *65*, 2479–2488. [CrossRef]
22. Xiao, Y.; Lu, L.Y.Y.; Liu, J.S.; Zhou, Z. Knowledge diffusion path analysis of data quality literature: A main path analysis. *J. Informetr.* **2014**, *8*, 594–605. [CrossRef]
23. ISO (International Standard Organization). *TECHNICAL REPORT ISO/TR 14049 Environmental Management—Life Cycle Assessment—Examples of Application of ISO 14041 to Goal and Scope Definition and Scope Definition and Inventory Analysis*; ISO: Geneva, Switzerland, 2006.

24. ISO (International Standard Organization). *ISO/TR 14049:2012. Environmental Management—Life Cycle Assessment—Illustrative Examples on How to Apply ISO 14044 to Goal and Scope Definition and Inventory Analysis*; ISO: Geneva, Switzerland, 2006.

25. ISO (International Standard Organization). *ISO 14041. Environmental Management—Life Cycle Assessment—Goal and Scope Definition and Inventory Analysis*; ISO: Geneva, Switzerland, 1998.

26. ISO (International Standard Organization). *ISO/TS 14072:2014. Environmental Management—Life Cycle Assessment—Requirements and Guidelines for Organizational Life Cycle Assessment*; ISO: Geneva, Switzerland, 2014.

27. ILCD. ILCD Handbook—General guide on LCA—Detailed guidance. *Constraints* **2010**, *15*, 524–525.

28. Manfredi, S.; Allacker, K.; Pelletier, N.; Chomkhamsri, K.; de Souza, D.M. *European Commission Product Environmental Footprint (PEF) Guide*; European Commission—Joint Research Centre: Ispra, Italy, 2012; p. 154.

29. European Commission. *PEFCR Guidance Document Guidance for the Development of Product Environmental Footprint Category Rules (PEFCRs)*, Version 6.3 ed; European Commission: Brussels, Belgium, 2017; p. 2018.

30. Moro, A.; Joanny, G.; Moretti, C. Emerging technologies in the renewable energy sector: A comparison of expert review with a text mining software. *Futures* **2020**, *117*, 102511. [CrossRef]

31. Batagelj, V.; Mrvar, A. Pajek—analysis and visualization of large networks. In *Graph Drawing Software*; Springer: Berlin/Heidelberg, Germany, 2011; ISBN 978-3-540-45848-7.

32. Ciano, M.P.; Strozzi, F.; Minelli, E.; Pozzi, R.; Rossi, T. The link between lean and human resource management or organizational behaviour: A bibliometric review. In Proceedings of the XXIV Summer School "Francesco Turco"—Industrial Systems Engineering, Brescia, Italy, 11–13 September 2019; pp. 321–328.

33. Strozzi, F.; Colicchia, C.; Creazza, A.; Noè, C. Literature review on the 'smart factory' concept using bibliometric tools. *Int. J. Prod. Res.* **2017**, *55*, 6572–6591. [CrossRef]

34. Liu, J.S.; Lu, L.Y.Y. An integrated approach for main path analysis: Development of the Hirsch index as an example. *J. Am. Soc. Inf. Sci. Technol.* **2012**, *63*, 528–542. [CrossRef]

35. Majeau-Bettez, G.; Dandres, T.; Pauliuk, S.; Wood, R.; Hertwich, E.; Samson, R.; Strømman, A.H. Choice of allocations and constructs for attributional or consequential life cycle assessment and input-output analysis. *J. Ind. Ecol.* **2018**, *22*, 656–670. [CrossRef]

36. De Camillis, C.; Brandão, M.; Zamagni, A.; Pennington, D. *Sustainability Assessment of Future-Oriented Scenarios: A Review of Data Modelling Approaches in Life Cycle Assessment. Towards Recommendations for Policy Making and Business Strategies*; Publications Office of the European Union: Luxemburg, 2013; ISBN 9789279325229.

37. Sandin, G.; Peters, G.M.; Svanström, M. Life cycle assessment of construction materials: The influence of assumptions in end-of-life modelling. *Int. J. Life Cycle Assess.* **2014**, *19*, 723–731. [CrossRef]

38. Peñaloza, D.; Erlandsson, M.; Falk, A. Exploring the climate impact effects of increased use of bio-based materials in buildings. *Constr. Build. Mater.* **2016**, *125*, 219–226. [CrossRef]

39. Heijungs, R. *Economic Drama and the Environmental Stage: Formal Derivation of Algorithmic Tools for Environmental Analysis and Decisionsupport from a Unified Epistemological Principle*; Leiden University: Leiden, The Netherlands, 1997.

40. Chen, C.; Habert, G.; Bouzidi, Y.; Jullien, A.; Ventura, A. LCA allocation procedure used as an incitative method for waste recycling: An application to mineral additions in concrete. *Resour. Conserv. Recycl.* **2010**, *54*, 1231–1240. [CrossRef]

41. Bailis, R.; Kavlak, G. Environmental implications of Jatropha biofuel from a silvi-pastoral production system in central-west Brazil. *Environ. Sci. Technol.* **2013**, *47*, 8042–8050. [CrossRef]

42. Schau, E.M.; Fet, A.M. LCA studies of food products as background for environmental product declarations. *Int. J. Life Cycle Assess.* **2008**, *13*, 255–264. [CrossRef]

43. Heijungs, R.; Guinée, J.B. Allocation and "what-if" scenarios in life cycle assessment of waste management systems. *Waste Manag.* **2007**, *27*, 997–1005. [CrossRef]

44. Heijungs, R. Ten easy lessons for good communication of LCA. *Int. J. Life Cycle Assess.* **2014**, *19*, 473–476. [CrossRef]

45. Marvuglia, A.; Cellura, M.; Heijungs, R. Toward a solution of allocation in life cycle inventories: The use of least-squares techniques. *Int. J. Life Cycle Assess.* **2010**, *15*, 1020–1040. [CrossRef]

46. Steubing, B.; Wernet, G.; Reinhard, J.; Bauer, C.; Moreno-Ruiz, E. The ecoinvent database version 3 (part II): Analyzing LCA results and comparison to version 2. *Int. J. Life Cycle Assess.* **2016**, *21*, 1269–1281. [CrossRef]

47. Nguyen, T.L.T.; Hermansen, J.E. System expansion for handling co-products in LCA of sugar cane bio-energy systems: GHG consequences of using molasses for ethanol production. *Appl. Energy* **2012**, *89*, 254–261. [CrossRef]

48. Corrado, S.; Ardente, F.; Sala, S.; Saouter, E. Modelling of food loss within life cycle assessment: From current practice towards a systematisation. *J. Clean. Prod.* **2017**, *140*, 847–859. [CrossRef]

49. Brander, M.; Wylie, C. The use of substitution in attributional life cycle assessment. *Greenh. Gas Meas. Manag.* **2011**, *1*, 161–166. [CrossRef]

50. Nhu, T.T.; Dewulf, J.; Serruys, P.; Huysveld, S.; Nguyen, C.V.; Sorgeloos, P.; Schaubroeck, T. Resource usage of integrated Pig-Biogas-Fish system: Partitioning and substitution within attributional life cycle assessment. *Resour. Conserv. Recycl.* **2015**, *102*, 27–38. [CrossRef]

51. Weidema, B. ISO System Expansion = Substitution 2.0. 2014. Available online: https://lca-net.com/blog/iso-system-expansion-substitution/ (accessed on 15 November 2019).

52. Forman, G.S.; Hahn, T.E.; Jensen, S.D. Greenhouse gas emission evaluation of the GTL pathway. *Environ. Sci. Technol.* **2011**, *45*, 9084–9092. [CrossRef]

53. Manninen, K.; Koskela, S.; Nuppunen, A.; Sorvari, J.; Nevalainen, O.; Siitonen, S. The applicability of the renewable energy directive calculation to assess the sustainability of biogas production. *Energy Policy* **2013**, *56*, 549–557. [CrossRef]

54. Karlsson, H.; Börjesson, P.; Hansson, P.A.; Ahlgren, S. Ethanol production in biorefineries using lignocellulosic feedstock—GHG performance, energy balance and implications of life cycle calculation methodology. *J. Clean. Prod.* **2014**, *83*, 420–427. [CrossRef]

55. García, C.A.; Fuentes, A.; Hennecke, A.; Riegelhaupt, E.; Manzini, F.; Masera, O. Life-cycle greenhouse gas emissions and energy balances of sugarcane ethanol production in Mexico. *Appl. Energy* **2011**, *88*, 2088–2097. [CrossRef]

56. Li, X.; Mupondwa, E. Life cycle assessment of camelina oil derived biodiesel and jet fuel in the Canadian Prairies. *Sci. Total Environ.* **2014**, *481*, 17–26. [CrossRef] [PubMed]

57. Brockmann, D.; Pradinaud, C.; Champenois, J.; Benoit, M.; Hélias, A. Environmental assessment of bioethanol from onshore grown green seaweed. *Biofuels Bioprod. Biorefining* **2015**, *9*, 696–708. [CrossRef]

58. Tillman, A.M.; Ekvall, T.; Baumann, H.; Rydberg, T. Choice of system boundaries in life cycle assessment. *J. Clean. Prod.* **1994**, *2*, 21–29. [CrossRef]

59. Giuntoli, J.; Commission, E.; Bulgheroni, C.; Commission, E.; Marelli, L.; Commission, E.; Sala, S.; Commission, E. Brief on the use of Life Cycle Assessment (LCA) to evaluate environmental impacts of the bioeconomy. *J. Sustain. Energy Environ. Spec. Issue* **2019**, 1–8. [CrossRef]

60. Cherubini, E.; Franco, D.; Zanghelini, G.M.; Soares, S.R. Uncertainty in LCA case study due to allocation approaches and life cycle impact assessment methods. *Int. J. Life Cycle Assess.* **2018**, *23*, 2055–2070. [CrossRef]

61. Herrmann, I.T.; Jørgensen, A.; Bruun, S.; Hauschild, M.Z. Potential for optimized production and use of rapeseed biodiesel. Based on a comprehensive real-time LCA case study in Denmark with multiple pathways. *Int. J. Life Cycle Assess.* **2013**, *18*, 418–430. [CrossRef]

62. Forman, G.S.; Hauser, A.B.; Adda, S.M. Life cycle analysis of gas to liquids (GTL) derived linear alkyl benzene. *J. Clean. Prod.* **2014**, *80*, 30–37. [CrossRef]

63. Weidema, B.P.; Frees, N.; Nielsen, A.M. Marginal production technologies for life cycle inventories. *Int. J. Life Cycle Assess.* **1999**, *4*, 48–56. [CrossRef]

64. Hermansson, F.; Janssen, M.; Svanström, M. Allocation in life cycle assessment of lignin. *Int. J. Life Cycle Assess.* **2020**. Article in Press. [CrossRef]

65. Pelletier, N.; Allacker, K.; Pant, R.; Manfredi, S. The European Commission Organisation Environmental Footprint method: Comparison with other methods, and rationales for key requirements. *Int. J. Life Cycle Assess.* **2014**, *19*, 387–404. [CrossRef]

66. Cherubini, F.; Strømman, A.H.; Ulgiati, S. Influence of allocation methods on the environmental performance of biorefinery products—A case study. *Resour. Conserv. Recycl.* **2011**, *55*, 1070–1077. [CrossRef]

67. Moretti, C.; Corona, B.; Rühlin, V.; Götz, T.; Junginger, M.; Brunner, T.; Obernberger, I.; Shen, L. Combining biomass gasification and solid oxid fuel cell for heat and power generation: An early-stage life cycle assessment. *Energies* **2020**, *13*, 2773. [CrossRef]

68. Ahlgren, S.; Björklund, A.; Ekman, A.; Karlsson, H.; Berlin, J.; Börjesson, P.; Ekvall, T.; Finnveden, G.; Janssen, M.; Strid, I. Review of methodological choices in LCA of biorefinery systems—Key issues and recommendations. *Biofuels Bioprod. Biorefining* **2015**, *9*, 606–619. [CrossRef]

69. Mackenzie, S.G.; Leinonen, I.; Kyriazakis, I. The need for co-product allocation in the life cycle assessment of agricultural systems—Is "biophysical" allocation progress? *Int. J. Life Cycle Assess.* **2017**, *22*, 128–137. [CrossRef]

70. Azapagic, A.; Clift, R. Allocation of environmental burdens in multiple-function systems. *J. Clean. Prod.* **1999**, *7*, 101–119. [CrossRef]

71. Azapagic, A.; Clift, R. Allocation of environmental burdens in co-product systems: Process and product-related burdens (part 2). *Int. J. Life Cycle Assess.* **2000**, *5*, 31–36. [CrossRef]

72. Ekvall, T.; Finnveden, G. Allocation in ISO 14041—A critical review. *J. Clean. Prod.* **2001**, *9*, 197–208. [CrossRef]

73. Azapagic, A.; Clift, R. Linear programming as a tool in life cycle assessment. *Int. J. Life Cycle Assess.* **1998**, *3*, 305–316. [CrossRef]

74. Bernier, E.; Lavigne, C.; Robidoux, P.Y. Life cycle assessment of kraft lignin for polymer applications. *Int. J. Life Cycle Assess.* **2013**, *18*, 520–5288. [CrossRef]

75. González-García, S.; Moreira, M.T.; Feijoo, G. Environmental performance of lignocellulosic bioethanol production from alfalfa stems. *Biofuels Bioprod. Biorefining* **2010**, *4*, 118–131. [CrossRef]

76. Finnveden, G.; Albertsson, A.C.; Berendson, J.; Eriksson, E.; Höglund, L.O.; Karlsson, S.; Sundqvist, J.O. Solid waste treatment within the framework of life-cycle assessment. *J. Clean. Prod.* **1995**, *3*, 189–199. [CrossRef]

77. Jungmeier, G.; Werner, F.; Jarnehammar, A.; Hohenthal, C.; Richter, K. Allocation in LCA of wood-based products—Experiences of cost action E9: Part II. Examples. *Int. J. Life Cycle Assess.* **2002**, *7*, 290–294. [CrossRef]

78. Nejad, A.T.M.; Saint-Antonin, V. Factors driving refinery CO_2 intensity, with allocation into products: Comment. *Int. J. Life Cycle Assess.* **2014**, *19*, 24–28.

79. Moretti, C.; Moro, A.; Edwards, R.; Rocco, M.V.; Colombo, E. Analysis of standard and innovative methods for allocating upstream and refinery GHG emissions to oil products. *Appl. Energy* **2017**, *206*, 372–381. [CrossRef]

80. International Dairy Federation. A Common Carbon Footprint Approach for Dairy: The IDF Guide to Standard Lifecycle Assessment Methodology for the Dairy Sector. 2015. Available online: https://store.fil-idf.org/product/a-common-carbon-footprint-approach-for-the-dairy-sector-the-idf-guide-to-standard-life-cycle-assessment-methodology/ (accessed on 10 July 2020).

81. Van der Harst, E.; Potting, J.; Kroeze, C. Comparison of different methods to include recycling in LCAs of aluminium cans and disposable polystyrene cups. *Waste Manag.* **2016**, *48*, 565–583. [CrossRef] [PubMed]

82. Silva, D.A.L.; Lahr, F.A.R.; Pavan, A.L.R.; Saavedra, Y.M.B.; Mendes, N.C.; Sousa, S.R.; Sanches, R.; Ometto, A.R. Do wood-based panels made with agro-industrial residues provide environmentally benign alternatives? An LCA case study of sugarcane bagasse addition to particle board manufacturing. *Int. J. Life Cycle Assess.* **2014**, *19*, 1767–1778. [CrossRef]

83. Palmieri, N.; Forleo, M.B.; Giannoccaro, G.; Suardi, A. Environmental impact of cereal straw management: An on-farm assessment. *J. Clean. Prod.* **2017**, *142*, 2950–2964. [CrossRef]

84. Vergé, X.; Maxime, D.; Desjardins, R.L.; Vanderzaag, A.C. Allocation factors and issues in agricultural carbon footprint: A case study of the Canadian pork industry. *J. Clean. Prod.* **2016**, *113*, 587–595. [CrossRef]

85. Tufvesson, L.M.; Tufvesson, P.; Woodley, J.M.; Börjesson, P. Life cycle assessment in green chemistry: Overview of key parameters and methodological concerns. *Int. J. Life Cycle Assess.* **2013**, *18*, 431–444. [CrossRef]

86. Vidal, R.; Martínez, P.; Garraín, D. Life cycle assessment of composite materials made of recycled thermoplastics combined with rice husks and cotton linters. *Int. J. Life Cycle Assess.* **2009**, *14*, 73–82. [CrossRef]

87. Vera, I.; Hoefnagels, R.; van der Kooij, A.; Moretti, C.; Junginger, M. A carbon footprint assessment of multi-output biorefineries with international biomass supply: A case study for the Netherlands. *Biofuels Bioprod. Biorefining* **2020**, *14*, 198–224. [CrossRef]

88. Ekvall, T.; Tillman, A.M. Open-loop recycling: Criteria for allocation procedures. *Int. J. Life Cycle Assess.* **1997**, *2*, 155–162. [CrossRef]

89. Clift, R. Chairman's report of session 3: Causality and allocation procedures. In Proceedings of the European Workshop on Allocation in LCA, Leiden, The Netherlands, 24–25 February 1994; pp. 3–4.

90. Clift, R. Report from setac-europe working group on life cycle inventory analysis. In *Abstract Book, 6th SETAC-Europe, Proceedings of the Annual Meeting, Taormina, Italy, 19–22 May 1996*; SETAC-Europe: Brussels, Belgium, 1996; p. 17.

91. Azapagic, A.; Clift, R. Life cycle assessment and linear programming environmental optimisation of product system. *Comput. Chem. Eng.* **1995**, *19*, 229–234. [CrossRef]

92. ISO (International Standard Organization). *ISO/TC 207/SC 5/AXIG2: CD 14 041.2. N99, DIN. 1996 (as Cited by Ekvall and Tillman in Open-loop Recycling: Criteria for Allocation Procedures-[88])*; ISO: Geneva, Switzerland, 1996.

93. Azapagic, A.; Clift, R. Allocation of Environmental Burdens in Co-product Systems: Product-related Burdens (Part 1). *Int. J. Life Cycle Assess.* **1999**, *4*, 357–369. [CrossRef]

94. Ekvall, T.; Weidema, B.P. System boundaries and input data in consequential life cycle inventory analysis. *Int. J. Life Cycle Assess.* **2004**, *9*, 161–171. [CrossRef]

95. Thrane, M. LCA of Danish fish products: New methods and insights. *Int. J. Life Cycle Assess.* **2006**, *11*, 66–74. [CrossRef]

96. Schmidt, J.H.; Weidema, B.P. Shift in the marginal supply of vegetable oil. *Int. J. Life Cycle Assess.* **2008**, *13*, 235–239. [CrossRef]

97. Dalgaard, R.; Schmidt, J.; Halberg, N.; Christensen, P.; Thrane, M.; Pengue, W.A. LCA for soybean meal. *LCA Food Prod.* **2008**, *10*, 240–254. [CrossRef]

98. Thomassen, M.A.; Dalgaard, R.; Heijungs, R.; De Boer, I. Attributional and consequential LCA of milk production. *Int. J. Life Cycle Assess.* **2008**, *13*, 339–349. [CrossRef]

99. Finnveden, G.; Hauschild, M.Z.; Ekvall, T.; Guinée, J.; Heijungs, R.; Hellweg, S.; Koehler, A.; Pennington, D.; Suh, S. Recent developments in Life Cycle Assessment. *J. Environ. Manag.* **2009**, *91*, 1–21. [CrossRef] [PubMed]

100. Bier, J.M.; Verbeek, C.J.R.; Lay, M.C. An eco-profile of thermoplastic protein derived from blood meal Part 1: Allocation issues. *Int. J. Life Cycle Assess.* **2012**, *17*, 208–219. [CrossRef]

101. Wardenaar, T.; Van Ruijven, T.; Beltran, A.M.; Vad, K.; Guinée, J.; Heijungs, R. Differences between LCA for analysis and LCA for policy: A case study on the consequences of allocation choices in bio-energy policies. *Int. J. Life Cycle Assess.* **2012**, *17*, 1059–1067. [CrossRef]

102. Van der Werf, H.M.G.; Nguyen, T.T.H. Construction cost of plant compounds provides a physical relationship for co-product allocation in life cycle assessment. *Int. J. Life Cycle Assess.* **2015**, *20*, 777–784. [CrossRef]

103. Pradel, M.; Aissani, L.; Canler, J.P.; Roux, J.C.; Villot, J.; Baudez, J.C.; Laforest, V. Constructing an allocation factor based on product- and process-related parameters to assess environmental burdens of producing value-added sludge-based products. *J. Clean. Prod.* **2018**, *171*, 1546–1557. [CrossRef]

104. Pradel, M.; Aissani, L. Environmental impacts of phosphorus recovery from a "product" Life Cycle Assessment perspective: Allocating burdens of wastewater treatment in the production of sludge-based phosphate fertilizers. *Sci. Total Environ.* **2019**, *656*, 55–69. [CrossRef] [PubMed]

105. *European Union Directive 2008/98/EC of the European Parliament and of the Council of 19 November 2008 on Waste and Repealing Certain Directives*; LexUriServ; European Union: Brussels, Belgium, 2008.

106. Curran, M.A.; Mann, M.; Norris, G. The international workshop on electricity data for life cycle inventories. *J. Clean. Prod.* **2005**, *13*, 853–862. [CrossRef]

107. Plevin, R.J.; Delucchi, M.A.; Creutzig, F. Using Attributional Life Cycle Assessment to Estimate Climate-Change Mitigation Benefits Misleads Policy Makers. *J. Ind. Ecol.* **2014**, *18*, 73–83. [CrossRef]

 © 2020 by the authors. Licensee MDPI, Basel, Switzerland. This article is an open access article distributed under the terms and conditions of the Creative Commons Attribution (CC BY) license (http://creativecommons.org/licenses/by/4.0/).

 *energies*

Article

Private and Externality Costs and Benefits of Recycling Crystalline Silicon (c-Si) Photovoltaic Panels

Elizabeth Markert [1], Ilke Celik [2] and Defne Apul [1],*

[1] Department of Civil and Environmental Engineering, University of Toledo, Toledo Lucas County, OH 43606, USA; Elizabeth.Markert@rockets.utoledo.edu

[2] Department of Electrical and Computer Engineering, University of Wisconsin–Platteville, Platteville Grant County, WI 53818, USA; celiki@uwplatt.edu

* Correspondence: Defne.Apul@utoledo.edu

Received: 15 May 2020; Accepted: 12 July 2020; Published: 15 July 2020

Abstract: With solar photovoltaics (PV) playing an increasing role in our global energy market, it is now timely and critical to understand the end of life management of the solar panels. Recycling the panels can be an important pathway, possibly recovering a considerable amount of materials and adding economic benefits from currently installed solar panels. Yet, to date, the costs and benefits of recycling, especially when externality costs resulting from environmental pollution are considered, are largely unknown. In this study, we quantified the private and externality costs and benefits of recycling crystalline silicon (c-Si) PV panels. We found that the private cost of end-of-life (EoL) management of the c-Si PV module is USD 6.7/m^2 and much of this cost is from transporting (USD 3.3/m^2) and landfilling (USD 3.1/m^2), while the actual recycling process (the cost of consumed materials, electricity or the investment for the recycling facilities) is very small (USD 0.3/m^2). We found that the external cost of PV EoL management is very similar to the private cost (USD 5.2/m^2). Unlike the breakdown of the private costs, much of the externality costs (USD 4.08/m^2) come from the recycling process, which suggests that more environmentally friendly methods (e.g., recycling methods that involve fewer toxic chemicals, acids, etc.) should be preferred. We estimated that the total economic value of the recycled materials from c-Si PV waste is USD 13.6/m^2. This means that when externality costs are not considered, the net benefit of recycling is USD 6.7; when the externality cost of recycling is considered, there is still a net benefit of USD 1.19 per m^2.

Keywords: end of life of PV; cost of PV recycling; photovoltaic waste; FRELP

---

1. Introduction

Solar photovoltaic (PV) technology plays an increasingly important role as a key energy source [1,2] As this technology grows, it is important to ensure that each process in the life cycle of PVs is sustainable [3,4]. The environmental impacts from manufacturing and operation of solar PV panels have been widely studied [5,6] and more recently, there has been a growing interest in understanding the environmental impacts of the end-of-life (EoL) management of solar panels [7–9]. Solar panels last from 20–30 years before weather and external conditions necessitate their retirement [10,11]. Because the mainstream, large-scale use of PV technology is relatively new, the infrastructure to recycle solar panels is not yet built for the capacity it must handle in the future [12]. The decrease in the price of PV modules' and the reduction in the environmental impact of solar systems in comparison to traditional fossil fuel technologies has led to many more large-scale solar plants being installed [11,13]. The global annual PV power capacity installed was equal to 114 GW in 2019, a net year-on-year increase of 17.5% from 2018 [14,15]. This rapid increase in panel use necessitates responsible, industrial-scale recycling and disposal processes.

In developing recycling processes for solar panels, it is important for us to understand both the cost and environmental impacts of the technology. The environmental impacts of EoL management of solar PV panels has received great attention recently; many authors have estimated the environmental impacts of EoL of solar panels using the life cycle assessment method [16–20] These studies highlighted that the majority of the impacts are associated with chemical usage for process recycling as well as the transportation of PV waste at the EoL of PVs. These studies also widely found that the environmental harm that can be avoided by recovering materials from PV panels is greater than the environmental harm caused from the energy and fuel that it takes to recycle them. The cost of recycling solar panels has also received great attention [21–24]. The cost assessment studies concluded that the PV recycling process cost—more specifically, using mechanical and/or thermal methods—is the major contributor to the cost of EoL management of PV waste. Since these studies either focus on the cost or environmental impacts, and since each study focuses on a different recycling process, we cannot deduce how cost and environmental information relate to one another for a given recycling process. Besides, environmental cost PV EoL management has been largely ignored in the literature.

In this study, we addressed this issue by studying the "full recovery end of life photovoltaic project" (FRELP) method of recycling crystalline silicone (c-Si) panels. The FRELP approach has found technological solutions for every step of the c-Si PV treatment process and has been successful in translating this information into a technically and economically feasible industrial process design [20]. We monetized the environmental impacts of c-Si PV recycling using this approach to obtain the externality costs of the FRELP method. We also analyzed the FRELP method to estimate its private costs. We then created our cost model, which compares these private to externality costs and showed the net benefits of the FRELP method by comparing the economic and environmental benefits that can be supplied from the recovered materials and processes associated with EoL management.

2. Method

2.1. Data Collection

There are several methods that have been proposed or are in use for the recycling of solar panels [21,22,25,26]. Of these, we followed the FRELP method, proposed by Latunussa et al., because, unlike other studies, FRELP was developed for industrial-scale recycling and all the details of life cycle inventories, including emissions from the facility and the efficiency of recycling, achieved yields data that were provided clearly, making the method easy to understand and utilize to expand for further cost analysis [20]. The FRELP method aims to test new methods and technologies with the goal of 100% recycling of PV panels in an economically responsible manner. Data on process costs and investment costs of the PV recycling process were developed using literature and industry data which can be found throughout the Supplementary Information tables (Tables S1 and S2).

The method of recycling we modeled by Latunussa et al., 2016, is acid leaching and electrolysis (Figure 1). In this process, after the PV panels are unloaded into the recycling facility, they are disassembled, the glass is separated and refined, and the PV sandwiches are cut and incinerated. The bottom ash from the incinerator is shipped to a different facility to be sent through several processes including sieving, acid leaching, filtration, electrolysis, neutralization, and a filter press. These processes for treating the bottom ash make up all the material inputs in the process cost of PV recycling and more than half of the electricity input. From these recycling steps, the recovery materials of aluminum scrap, silicon scrap, silver scrap, copper scrap, and glass scrap are recovered. Liquid wastes, sludge, hazardous fly ash, and contaminated glass produced during the entire recycling process (all waste boxes in Figure 1) are sent to landfills.

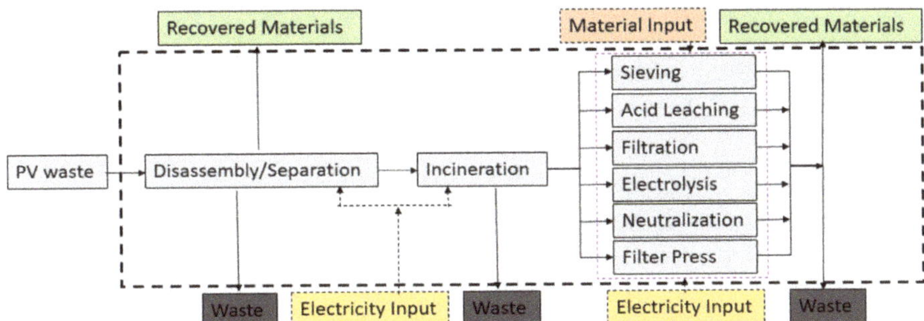

Figure 1. Simplified PV recycling process modeled from Latunussa et al., 2016 [20].

Figure 2 shows the costs and benefits considered in the analysis. The data were broken down into several different components: process costs, investment costs, environmental externality costs, recovered material costs, transportation costs, policy benefit costs, and landfilling tipping costs. The functional unit used in this work is 1 m^2 of PV. The private costs (defined as the market cost for a technology or production), external costs, and benefit costs are all broken down in terms of this unit.

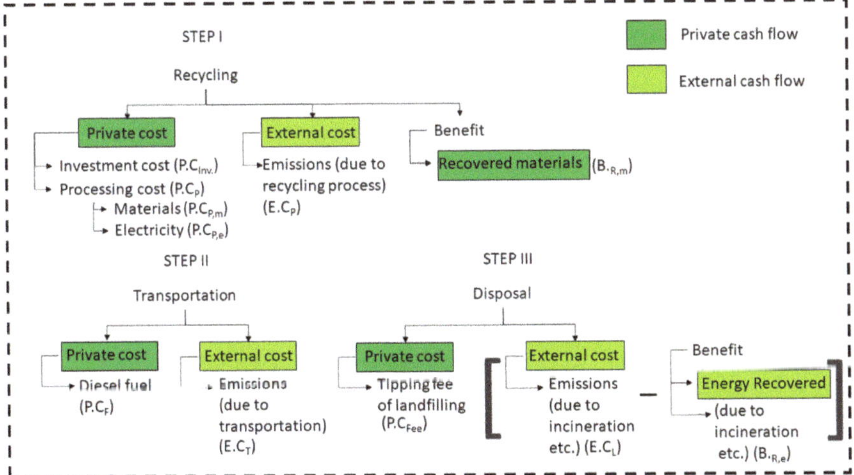

Figure 2. Our framework in analyzing the cost of FRELP recycling method.

To quantify the different values shown in Figure 2, the below equations were used:

$$\text{Total cost of PV Recycling} = \sum \text{Private Cost} + \sum \text{External Cost} - \sum \text{Benefits} \tag{1}$$

$$\text{Private cost of PV recycling} = \text{P.C}_{Inv} + \text{P.C}_{P,m} + \text{P.C}_{P,e} + \text{P.C}_F + \text{P.C}_{Fee} \tag{2}$$

$$\text{External cost of PV recycling} = \text{E.C}_P + \text{E.C}_T + \text{E.C}_L \tag{3}$$

$$\text{Benefits of PV recycling} = \text{B.}_{R,e} + \text{B.}_{R,m.} \tag{4}$$

In calculating the private cost (P.C) values, we consider all the private transactions that a PV recycler would have to pay during the end of life management of PV waste. For example, throughout the recycling process of PV waste (STEP I in Figure 2), the PV recycler must invest in purchasing the instruments (See Supplementary Information Table S2 required for operating the process (P.C$_{Inv}$)

(see Table S2). In addition, the cost of materials (P.C$_M$) (see Table S1) and electricity (P.C$_E$) (Table S1) utilized in operating these instruments are considered. Similarly, during the transportation (STEP II in Figure 1) of PV waste (from installation location to recycling center and from recycling center to landfills), private cash flows are associated with the cost of fuel consumption of the trucks (P.C$_F$) (Table S3) that carries the PV waste. Finally, during the landfilling of unrecovered materials (STEP III in Figure 1), a tipping fee is paid for dumping non-hazardous materials, P.C$_{Fee}$ (Table S4).

The external cost in Equantion 1 refers to the cost of environmental damage resulting from the pollutants released during the end-of-life management of PVs, expressed as the dollar value. The external cost of the emissions from recycling process (E.C$_P$) (Table S5), transportation (E.C$_T$) (Table S6), and incineration during the landfilling (E.C$_L$) (see Table S7) are considered in this analysis.

Equations (2)–(4) show all the components of our cost calculation framework. Note that we normalized all these values in terms of 1 m^2 c-Si PV module, since a unit surface area is the functional unit of this study.

Finally, the economic value of benefits of end-of-life management associated with the recycling and landfilling of PV waste were assessed. The private benefit value of recovered materials (B$_{R,m}$ e.g., from scrap metals) (Table S8) and energy (B$_{R,e}$ e.g., as electricity) (Table S7) are analyzed as positive private cash flows in the framework.

2.2. Private Cost

P.C$_{Inv}$ is important to discuss in the private cost assessment of PV recycling because, before being able to make a turn of profit on PV recycling, an infrastructure must be set up. This can require significant amounts of money as many innovative PV recycling processes warrant new, specialized equipment. Another issue is that the profitability of investments related to the construction of PV recycling facilities and equipment is guaranteed only by the management of great amounts of wastes [23]. In small or lab-scale operations, high investment costs may mean that a facility never turns a profit. This paper draws its investment costs from processes required in the FRELP method, in which insight into the steps of the PV recycling techniques was given. The costs of equipment were taken from manufacturers' websites [27,28]. In order to complete the recycling processes described in the paper the costs of purchasing, these technologies must be taken into account. Although these are traditionally one-time costs, the lifetime of machinery, yearly mass produced, and number of panels in our functional unit were considered to represent the investment cost of PV relative to the other process costs on a 1 m^2 basis. Table S2 in the Supplementary Information tables is a collection of the supplies and costs of the equipment needed for c-Si investment for the given recycling process.

The transportation costs were found by utilizing Latunussa data on distance traveled (km), averaging diesel cost, and estimating average semi-truck fuel efficiency. This can be found in Supplementary Information Table S3.

2.3. External Cost

The environmental externality cost (E.C) data were estimated by multiplying the emissions from the recycling process with the damage cost per mass of emissions

$$E.C = \text{Emission (kg)} \times \text{Damage Cost (USD/kg)}. \tag{5}$$

Emission data were taken directly from Latunussa's paper. The PV recycling pollutant numbers were multiplied by the found environmental externality multipliers to determine the approximate cost of the process' impact per material. The data for these externality values comes from four sources [23,29–32] The Supplementary Information tables include all externality tables for the recycling of PV panels (Tables S5–S7). Below are the impact categories used in this paper and their corresponding damage cost values. Within the four papers from which this data was pulled, some variation occurred between values. For simplicity in this paper, the damage cost number listed below was pulled from its most recent publicized representation. For example, if a value for CED was found in three of the four papers, the value used in this paper would be pulled from the most recently published work. The nature of the gaps between the values can be investigated in their original work, but, in short, different LCA metrics have different environmental impacts which cause distinct environmental damage. Some of the metrics with higher damage costs are those which have impacts that society deems more disruptive and unsafe than others.

2.4. Benefits

The benefit cost data were found by adding the monetized energy recovered from the incineration disposal process and the cost payback from the materials recovered during the recycling process

$$\text{Benefits of PV recycling} = B_{\cdot R,e} + B_{\cdot R,m}. \tag{6}$$

Recovery yield was included within the scope of this paper because it is an important aspect when evaluating the success of the cost-saving measures of using non-virgin materials. Although the recycling process requires high up-front investment costs, money can be saved in the purchasing of new materials. This is because many of these materials are saved after the recycling process and can be sold to manufacturers or, in the case of First Solar, with its identity as both a manufacturing and recycling facility, processed again as recycled feed (First Solar's Module Collection and Recycling Program). The beneficiary of the recovered products depends on whether there is a closed-loop vs. open-loop process. In an open-loop process, materials are sold to external processing facilities, and in a closed-loop process, the materials are sent to make more of the same product they had been pre-recycled.

In this work the percent yields of recovered materials were taken from the literature [20]. Using these numbers, the cost of the recovered materials per m^2 were found. Recovery cost tables can be found in the Supplementary Information tables (Tables S5 and S6).

3. Results

3.1. Private Cost of Recycling

The total EoL cost of 1 m^2 of c-Si PV module was found to be USD 6.72/m^2 (Figure 3). Of the three cost components of PV EoL management, the transportation-associated cost was found to be the highest (USD 3.36/m^2) while the cost of the recycling process (the cost of consumed materials, electricity or the investment for the recycling facilities) was found to be the most insignificant (USD 0.25/m^2).

The main reason for the high transportation cost can be attributed to the long distance of transportation required in the end-of-use of PV panels. Figure 3 also shows each transportation step and its cost in carrying them with a truck, as referred to in the FRELP method. The costs of each transportation step correlate directly with the distance between facilities. As can be derived from Figure 3, the long distances from deployment locations to PV recycling facilities are the main drivers of the high cost due to transportation in the EoL of PVs (an average of 400 km away from the collection points [20]). The table containing these distances can be found in Supplementary Information Table S3 (note that the cost of transportation is linearly proportional with traveled distances).

The disposal cost of c-Si PV waste is made up of the tipping fees for four materials such as contaminated glass, fly ash, liquid waste and sludge (Figure 3) in landfills. The landfilling associated cost can be attributed to the high cost of tipping fees for the sludge treatment that consists of hazardous materials. The recycling of 1 m^2 of c-Si PV resulted in the generation of 0.70 kg of sludge, which equals approximately 90% of the total landfilling cost of c-Si PV's EoL management. For every 1000 kg of PV panels processed, 374.4 kg of waste is produced [20].

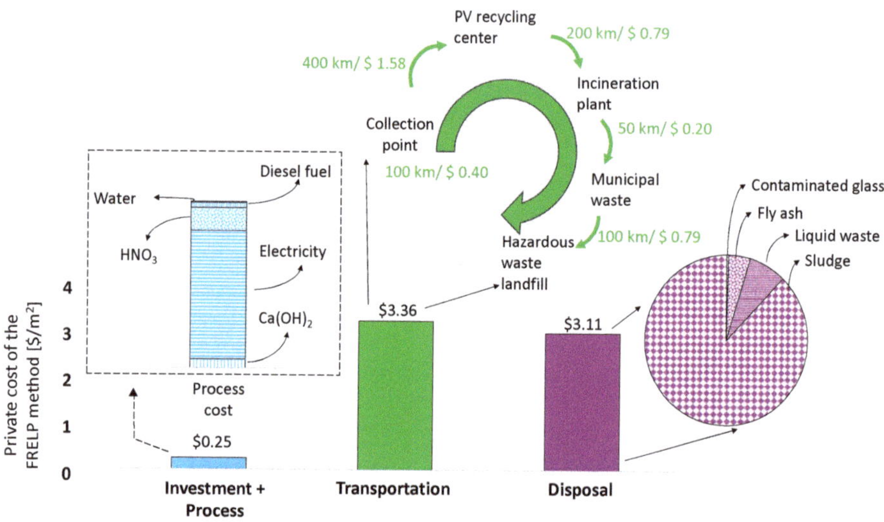

Figure 3. The end of life (EoL) management (private) cost of EoL of 1 m^2 of c-Si PV.

The total cost of the investment and processing for recycling has a minor impact on the private cost of c-Si PV's EoL management. In the FRELP recycling method, diesel fuel, electricity, nitric acid (HNO$_3$), water, and calcium hydroxide (Ca(OH)$_2$) were consumed during the recycling process of c-Si PV modules. Figure 3 also shows the cost breakdown of these process inputs. As seen, the electricity consumed is the most expensive cost component of the PV recycling process, with a value of 12 cents/kWh. Electricity is only used in the recycling process, totaling 1.55 KWh. The electricity is consumed in multiple steps during recycling such as during disassembly of the module, glass separation, and PV sandwich cutting. The second-largest cost component of the recycling process can be contributed to the chemicals required. The FRELP method utilizes 0.1 kg of nitric acid per 1 m^2 during the acid leaching process. Another chemical utilized during the FRELP method is calcium hydroxide, which is needed for the neutralization step of PV dismantling. This step is necessary because of the need to neutralize the nitric acid used in the acid leaching step beforehand. A total of 0.5 kg of calcium hydroxide per 1 m^2 is used, making it the largest amount of material required for the direct PV dismantling process, disregarding water. We found the impact of the investment cost for the instruments required for the recycling of c-Si PV panels is insignificant (¢1.1/m^2). The most outstanding cost component among the eight pieces of the instruments required for the FRELP process was found in the cartesian robot system, making up ¢0.6/m^2. Note that the instruments used in PV recycling can process about 8 million kg of PV waste annually and they can be used for long period of time (approximately 20 years) [20]. Therefore, their impact on processing 1 m^2 of the PV module was found to be very limited.

Faircloth et al. [33] also calculated the processing cost of the FRELP method as USD 0.03 per kg PV waste, which equals USD 0.48 per m^2 of PV waste, assuming that the FRELP recycling facility operates in Thailand. The difference (24 cents per m^2) between the processing cost of our and Faircloth et al.'s [33] results can be attributed to the differences in the cost of materials (acids, diesel fuel, water etc.) and electricity in the US and Thailand which are required for the FRELP process, as well as differences in methods of recycling. Similarly, we compared the disposal cost of the unrecovered materials from the recycling of PV waste with Faircloth et al.'s values [33]. However, the high cost of transportation was not identified in the literature. The reason could be the difference in the modeling approaches on transportation distances. Faircloth et al [33] assumed the total distance of travel as 100 km while Latunnusa et al [20] reported the overall distance of travel as 850 km. The type of fuel used is another factor which could cause discrepancies between different papers about PV transportation. However, diesel fuel is used by both this paper and Faircloth et al. [33].

3.2. External Cost of Recycling

We found that the total external cost of PV EoL management is about USD 5.7/m^2. The PV recycling process makes up the majority of the external costs (USD 4.08/m^2) since the environmental impacts associated with the processing of PV waste have a greater environmental impact than the sum of both the impacts of transportation (USD 0.73/m^2) and landfilling (USD 0.36/m^2). In fact, both transpiration and landfilling combined only make up 19% of the total external cost of PV EoL management. Various reasons cause this contrast. First off, the landfilling value is low because the positive impact of recovered energy from landfill incineration is factored into the total. This energy recovered offsets some of the negative externality costs of landfilling. The recycling process uses different chemicals such as HNO_3 and $Ca(OH)_2$, which are particularly harmful for the environment. Both the transportation and recycling processes require the input of diesel fuel, however, because the recycling process requires other chemical inputs as well, it acquires a higher external cost. This relationship can be seen in Figure 4 below.

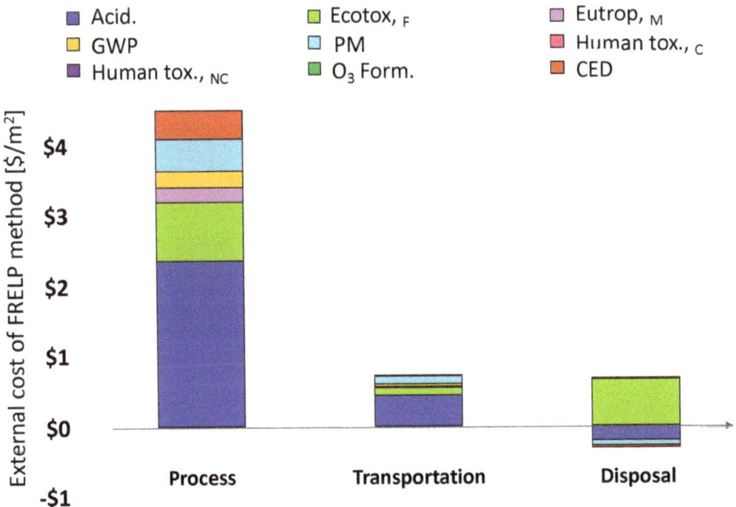

Figure 4. External costs of solar photovoltaic (PV) recycling.

The emissions associated with acidification, ecotoxicity and particulate matter (PM) impacts have the greatest external cost. Particularly, acidification-related emissions dominate the recycling and transportation steps of c-Si PVs' EoL phases. About 75% of acidifying emissions of the recycling process are released to the environment during the sieving and acid leaching of c-Si modules to separate the content of the materials of the silicon wafer and during the filtration, electrolysis, and neutralization processes to extract copper, metallurgical grade silicon and silver materials from the silicon wafer [20]. The acidifying emissions of transportation are due to the upstream emissions of diesel fuel extraction in refineries [20]. Figure 4 also shows that the external cost of landfilling is dominated by the particulate matter (PM) emissions. These PMs are released during the incinerations of polymer-based materials such as cables used in the balance-of-system of PV modules and encapsulant materials (Ethylene Vinyl Acetate (EVA)). Note that another reason that acidification and particulate matters dominate the external cost of c-Si PV's EoL management is that these two impact categories are priced with the highest externality cost according to x and y sources, with USD 341.80 kg $PM_{2.5\text{-eq}}$ and USD 64.47 per kg of SO_2, for particulate matter and acidification impacts, respectively.

The second highest external cost contributor to the recycling process and transportation steps of c-Si PV's EoL management was found to be the freshwater ecotoxicity (Ecotox.,$_F$). This is important to note because the cost value associated with freshwater ecotoxicity is on the lower end of the damage cost values shown in Table 1, which indicates that the number of ecotoxic emissions is greater than other emissions during the EoL management of PVs. Similar to particulate matter impacts, the incinerations of polymer-based materials are found to be the reasons for the harmful ecotoxic emissions to freshwater. The incineration process does not involve direct emissions to water. Rabl et al. explain that the air emissions from incinerators end up in water bodies due to the wet or dry deposition of aerosols [34]. Among the ecotoxic emissions from incinerators, the impact of mercury, lead and chromium was found to be dominant [34].

Table 1. Damage cost of different impact categories.

Impact Category [20]	Abbreviation	Damage Cost ($/kg)
Cumulative Energy Demand [MJ]	CED	0.01 [29]
Freshwater ecotoxicity [CTU$_e$]	Ecotox.,$_F$	0.05 [30]
Marine eutrophication [kg N$_{eq}$]	Eutrop.,$_M$	14.18 [30]
Freshwater eutrophication [kg P$_{eq}$]	Eutrop.,$_F$	2.02 [30]
Acidification [molc H+$_{eq}$]	Acid.	64.47 [31]
Photochemical ozone formation [NMVOC$_{eq}$]	O_3 Form.	2.33 [31]
Particulate matter [kg PM$_{2.5\,eq}$]	PM	341.80 [31]
Human toxicity, non-cancer effects [CTU$_h$]	Human tox., $_{NC}$	0.02 [30]
Human toxicity, cancer effects [CTU$_h$]	Human tox., $_C$	0.02 [30]
Global warming potential [kg CO$_{2\text{-eq}}$]	GWP	0.04 [31]

The net external cost of disposal was found to be the lowest. In fact, the external cost of the landfilling process is very similar to transportation; however, the negative cost (avoided impacts) due to heat generation in the incineration creates credit values, which resulted in a lower external cost during the landfilling.

3.3. Cost–Benefit Analysis of Recycling

Figure 5 offers a cost–benefit analysis on recycling c-Si PV modules in their EoL. The negative cost values (benefits) show the revenue that can be realized the PV waste, while the positive cost values indicate the private and external cost of PV EoL management. The economic value of recovered materials through FRELP method was found to be USD 13.6/m^2 while the private cost of the method was calculated as USD 6.7/m^2. Taking the difference results in USD 7 of net economic benefit per recycling 1 m^2 of c-Si waste PV panel using FRELP method. This high net economic benefit value is due to the high material recovery rates in the FRELP. In the FRELP method, the material recovery rates are 96.2% (aluminum), 98% (glass), 74.5% (copper), 38.6% (silicon) and 11.3% (silver) [20]. However, the breakdown of the total revenue that can be realized from the reselling of the material recovery is 46% (aluminum), 25% (silver), 15% (glass), 11% (silicon) and 3% (copper), based on the market prices of these commodities [35] (see Supplementary Information Table S8). These results indicate that the substantial increases in the net economic benefit from c-Si PVs are hindered by the limited recovery rate of silver. For example, increasing the recovery rate of silver from 11% to 20% will result in a nearly 25% increase (from 13.62 to 16.22) in the total benefits from material recovery.

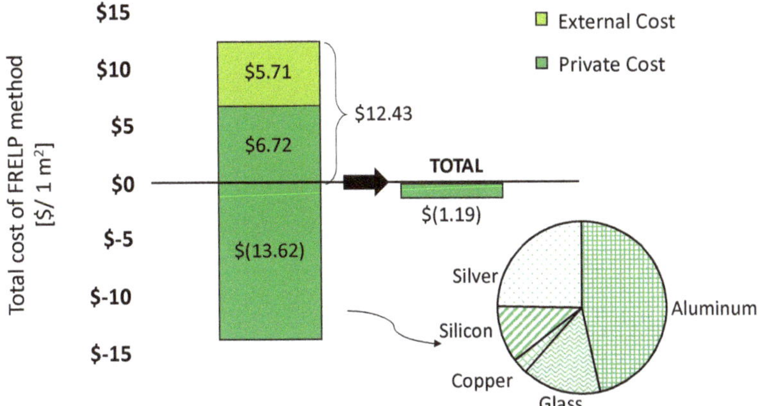

Figure 5. Total cost breakdown of the full recovery end-of-life photovoltaic project (FRELP) method.

The recycling process is a net benefit to the society, even when we include the externality costs of recycling. The total cost of c-Si recycling, including both private and external costs, is USD 12.43 without considering the benefit from recovered materials. When this benefit is added, the cost—or more accurately the benefit—is USD 1.19, meaning there is still a benefit from recycling c-Si PV even when the externality costs of recycling are considered. The total cost of PV recycling we found is USD −1.19/m^2, meaning that it is cheaper to recycle and use PV panels made from recycled materials than it is to throw these materials away at their end-of-life and use virgin materials.

Our net economic benefit results are consistent with the existing literature studies. Adamo et al. [36] also analyzed the cost of waste c-Si PV waste and reported that the cost of c-Si PV recycling varies from 8–19 EUR/m^2 (assuming that the c-Si PV module weighs 16 kg/m^2). The higher benefits in Adamo et al.'s result can be explained due to the modeling approach of transportation and disposal. Adamo et al. ignored the cost of transportation and tipping fees and only modeled the cost of the recycling process for the net economic benefit. If we performed the same analysis as Adamo, our result would show a USD 13/m^2 benefit, which is consistent with Adamo et al.'s result. Choi and Fthenakis calculated the net benefit as about 23 USD/module for thin-film CdTe PV technology [7,36]. The higher benefits from CdTe technology can be attributed to the high yields (>90%) of expensive materials (i.e., tellurium, cadmium) from the recycling of thin films.

3.4. Comparing the Cost of Virgin Materials to Recovered Materials

One final comparison which is important to note is the difference in the private and external costs for virgin materials vs. recovered materials, shown in Figure 6. In Figure 5, we estimate the value of the recovered materials as USD 13.62 but, in Figure 6, the cost of the virgin materials is about USD 90/m^2 [37]. There are two main reasons why the virgin material calculation and recovered material calculation are so different from one another. First, the recycling process is not able to recover all the valuable materials that go into making the PV panels, which makes the recovered material payback cost inevitably lower than the input (virgin) cost. Second, the virgin cost estimate includes the machinery and equipment needed for the extraction of materials, while the recovered material cost does not, as recovery does not bare these costs.

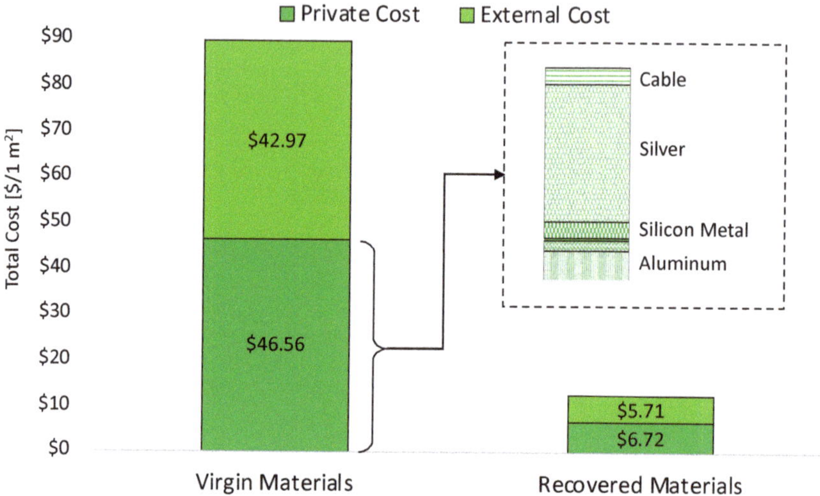

Figure 6. Virgin material vs. recovered material total cost. The virgin materials' private and external cost data were pulled from our group's earlier study [19], while the recovered materials' data were pulled from the estimations made in this paper.

There is a large difference in the two private cost values in Figure 6 because the cost of purchasing all the materials to construct 1 m² of the virgin PV panel is significantly more than the cost of purchasing the materials to recycle these panels. This is due to the fact that many of the metals needed to construct the crystalline silicon panels, such as silver, can be quite expensive. The external cost of using virgin materials is also higher, considering that these materials must be extracted from their original sources, which uses heavy equipment and transportation, while the external cost of recovering the materials comes exclusively from the discharge of chemicals and energy used to break down spent panels. Overall, it is clear that the use of virgin materials is cost-prohibitive as their recycled counterparts have funds of significantly less value. Recovered materials also include the cost savings of selling back these recovered products, which ultimately makes the recycling process a gain cost (or cost positive) for the recycler.

4. Conclusions

In this study, we calculated the private and externality cost of PV recycling in their EoL. We investigated the FRELP method that is used in recovering materials from c-Si PV waste. Our results indicate that the cost of EoL management of c-Si PV module are USD 6.72/m² and of the cost components of PV EoL management, the transportation associated cost was found to be the highest (USD 3.36/m²) while the cost of the recycling process (the cost of consumed materials, electricity or the investment for the recycling facilities) was found to be the most insignificant (USD 0.25/m²). We calculated that the external cost of PV EoL management is very similar to the private cost (about USD 5.2 m²). To the best of our knowledge, this is the first study to analyze the external cost of PV recycling. We found that the majority of the external costs of PV EoL management are associated with the processing of PV waste, USD 4.08/m² which suggests that more environmentally friendly methods (e.g., recycling methods that involve fewer toxic chemicals, acids, etc.) should be preferred in the recycling process of PV waste. Finally, we showed that the net economic benefit of FRELP method is almost USD 7 per m² module when external costs of recycling are not considered and USD 1.19 per m² when the external costs of recycling are considered. However, because this net benefit is so small, appropriate policies may need to be enacted to further motivate PV manufacturers to recycle their EoL panels.

Supplementary Materials: The following are available online at http://www.mdpi.com/1996-1073/13/14/3650/s1.

Author Contributions: E.M. is primarily responsible for data generation and writing the first draft of the manuscript. I.C. worked on interpretation of data and revisions to address Reviewers' comments. D.A. provided supervision and guidance. All authors have read and agreed to the published version of the manuscript.

Funding: This research was funded by the University of Toledo's Office of Undergraduate Research and SAIF grant provided by Office of Research and Sponsored Programs at the University of Wisconsin–Platteville.

Conflicts of Interest: The authors declare no conflict of interest.

Nomenclature

Abbreviation	Explanation
$P.C_{Inv}$	Private cost of investment
$P.C_P$	Private cost of process
$P.C_{P,m}$	Private cost of materials
$P.C_{P,e}$	Private cost of electricity
$E.C_P$	External cost of process
$B._{R,m}$	Benefit cost of recovered materials
$P.C_F$	Private cost of diesel fuel
$E.C_T$	External cost of transportation
$P.C_{Fee}$	Private cost of landfill tipping fee
$E.C_L$	External cost of incineration emissions
$B._{R,e}$	Benefit cost of recovered energy
CED	Cumulative Energy Demand
$Ecotox._{,F}$	Freshwater ecotoxicity
$Eutrop._{,M}$	Marine eutrophication
$Eutrop._{,F}$	Freshwater eutrophication
Acid.	Acidification
O_3 Form.	Photochemical ozone formation
PM	Particulate matter
Human tox., $_{NC}$	Human toxicity, non-cancer effects
Human tox., $_C$	Human toxicity, cancer effects
GWP	Global warming potential

References

1. Celik, I.; Mason, B.E.; Phillips, A.B.; Heben, M.J.; Apul, D. Environmental Impacts from photovoltaic solar cells made with single walled carbon nanotubes. *Environ. Sci. Technol.* **2017**, *51*, 4722–4732. [CrossRef] [PubMed]

2. Ahangharnejhad, R.H.; Phillips, A.B.; Ghimire, K.; Koirala, P.; Song, Z.; Barudi, H.M.; Barudi, A.; Sengupta, M.; Ellingson, R.J.; Yan, Y.; et al. Irradiance and temperature considerations in the design and deployment of high annual energy yield perovskite/CIGS tandems. *Sustain. Energy Fuels* **2019**, *3*, 1841–1851. [CrossRef]

3. Celik, I.; Philips, A.B.; Song, Z.; Yan, Y.; Ellingson, R.J.; Heben, M.J.; Apul, D. Energy Payback Time (EPBT) and Energy Return on Energy Invested (EROI) of perovskite tandem photovoltaic solar cells. *IEEE J. Photovolt.* **2017**, *8*, 2156–2161. [CrossRef]

4. Briese, E.; Piezer, K.; Celik, I.; Apul, D. Ecological network analysis of solar photovoltaic power generation systems emily. *J. Clean. Prod.* **2019**, *223*, 368–378. [CrossRef]

5. Celik, I.; Song, Z.; Cimaroli, A.; Yan, Y.; Heben, M.; Apul, D. Life Cycle Assessment (LCA) of perovskite PV cells projected from lab to fab. *Sol. Energy Mater. Sol. Cells* **2016**, *156*, 157–169. [CrossRef]

6. Celik, I.; Phillips, A.B.; Song, Z.; Yan, Y.; Ellingson, R.J.; Heben, M.J.; Apul, D. Environmental analysis of perovskites and other relevant solar cell technologies in a tandem configuration. *Energy Environ. Sci.* **2017**, *10*, 1874–1884. [CrossRef]

7. Lunardi, M.M.; Alvarez-Gaitan, J.P.; Bilbao, J.I.; Corkish, R. A review of recycling processes for photovoltaic modules. *Sol. Panels Photovolt. Mater.* **2018**. [CrossRef]

8.   Fiandra, V.; Sannino, L.; Andreozzi, C.; Graditi, G. End-of-life of silicon PV panels: A sustainable materials recovery process. *Waste Manag.* **2019**, *84*, 91–101. [CrossRef]

9.   Jung, B.; Park, J.; Seo, D.; Park, N. Sustainable system for raw-metal recovery from crystalline silicon solar panels: From noble-metal extraction to lead removal. *ACS Sustain. Chem. Eng.* **2016**, *4*, 4079–4083. [CrossRef]

10.   Wu, S.R.; Celik, I.; Apul, D.; Chen, J. A social impact quantification framework for the resource extraction industry. *Int. J. Life Cycle Assess.* **2019**, *24*, 1898–1910. [CrossRef]

11.   Celik, I.; Song, Z.; Phillips, A.B.; Heben, M.J.; Apul, D. Life cycle analysis of metals in emerging photovoltaic (PV) technologies: A modeling approach to estimate use phase leaching. *J. Clean. Prod.* **2018**, *186*, 632–639. [CrossRef]

12.   Deng, Y.; Liu, H.; Zhao, X.; Jiaqiang, E.; Chen, J. Effects of cold start control strategy on cold start performance of the diesel engine based on a comprehensive preheat diesel engine model. *Appl. Energy* **2018**, *201*, 279–287. [CrossRef]

13.   Beránek, V.; Olšan, T.; Libra, M.; Poulek, V.; Sedláček, J.; Dang, M.Q.; Tyukhov, I.I. New monitoring system for photovoltaic power plants' management. *Energies* **2018**, *11*, 2495. [CrossRef]

14.   Willuhn, M. *Global PV Market: 114 GW to Be Installed in 2019, with Continued Growth Onwards*; PV Magazine: Berlin, Germany, 2019.

15.   Celik, I. *Eco-Design of Emerging Photovoltaic (PV) Cells*; University of Toledo: Toledo, OH, USA, 2018.

16.   Giwa, A.S.; Xu, H.; Wu, J.; Li, Y.; Chang, F.; Zhang, X.; Jin, Z.; Huang, B.; Wang, K. Sustainable recycling of residues from the food waste (FW) composting plant via pyrolysis: Thermal characterization and kinetic studies. *J. Clean. Prod.* **2018**, *180*, 43–49. [CrossRef]

17.   Tao, J.; Yu, S. Review on feasible recycling pathways and technologies of solar photovoltaic modules. *Sol. Energy Mater. Sol. Cells* **2015**, *141*, 108–124. [CrossRef]

18.   Smith, Y.R.; Bogust, P. Review of solar silicon recycling. *Miner. Met. Mater. Ser. Part F* **2018**, *6*, 463–470.

19.   Maani, T.; Celik, I.; Heben, M.J.; Randall, J. Environmnetal impacts of recycling crystalline silicon (c-Si) and cadmium telluride (CdTe) solar panels. *Sci. Total Environ.* **2020**, *735*, 138827. [CrossRef]

20.   Latunussa, C.E.L.; Ardente, F.; Blengini, G.A.; Mancini, L. Life cycle assessment of an innovative recycling process for crystalline silicon photovoltaic panels. *Sol. Energy Mater. Sol. Cells* **2016**, *156*, 101–111. [CrossRef]

21.   Deng, R.; Chang, N.L.; Ouyang, Z.; Chong, C.M. A techno-economic review of silicon photovoltaic module recycling. *Renew. Sustain. Energy Rev.* **2019**, *109*, 532–550. [CrossRef]

22.   Choi, J.K.; Fthenakis, V. Design and optimization of photovoltaics recycling infrastructure. *Environ. Sci. Technol.* **2010**, *44*, 8678–8683. [CrossRef]

23.   Cucchiella, F.; D'Adamo, I.; Rosa, P. End-of-Life of used photovoltaic modules: A financial analysis. *Renew. Sustain. Energy Rev.* **2015**, *47*, 552–561. [CrossRef]

24.   di Nola, A.; Flondor, P.; Leuştean, I. MV-modules. *J. Algebr.* **2003**, *267*, 21–40. [CrossRef]

25.   Fthenakis, V.M. End-of-life management and recycling of PV modules. *Energy Policy* **2000**, *28*, 1051–1058. [CrossRef]

26.   Choi, J.-K.; Heo, J.-B.; Ban, S.-J.; Yi, S.-M.; Zoh, K.-D. Source apportionment of $PM_{2.5}$ at the coastal area in Korea. *Sci. Total Environ.* **2013**, *447*, 370–380. [CrossRef]

27.   Worx, R. How Much Do Industrial Robots Cost. Available online: https://www.robots.com/faq/how-much-do-industrial-robots-cost (accessed on 8 October 2019).

28.   Alibaba Manufacturers, Suppliers, Exporters & Importers. Available online: https://www.alibaba.com/?spm=a2700.details.scGlobalHomeHeader.9.df0b7eeeTwMxhR (accessed on 19 July 2019).

29.   Pizzol, M.; Weidema, B.; Brandao, M.; Osset, P. Monetary valuation in life cycle assessment: A review. *J. Clean. Prod.* **2015**, *86*, 170–179. [CrossRef]

30.   Keijzer, E. The environmental impact of activities after life: Life cycle assessment of funerals. *Int. J. Life Cycle Assess.* **2017**, *22*, 715–730. [CrossRef]

31.   Martinez-Sanchez, V.; Levis, J.W.; Damgaard, A.; DeCarolis, J.F.; Barlaz, M.A.; Astrup, T.F. Evaluation of externality costs in life-cycle optimization of municipal solid waste management systems. *Environ. Sci. Technol.* **2017**, *51*, 3119–3127. [CrossRef]

32.   Environmental Prices Handbook 2017 Methods and numbers for valuation of environmental impacts. 2018, pp. 1–177. Available online: https://www.cedelft.eu/en/publications/2113/envionmental-prices-handbook-2017 (accessed on 17 July 2019).

33. Faircloth, C.C.; Wagner, K.H.; Woodward, K.E.; Rakkwamsuk, P.; Gheewala, S.H. The environmental and economic impacts of photovoltaic waste management in Thailand. *Resour. Conserv. Recycl.* **2019**, *143*, 260–272. [CrossRef]

34. Rabl, A.; Spadaro, J.V.; Zoughaib, A. Environmental impacts and costs of solid waste: A comparison of landfill and incineration. *Waste Manag. Res.* **2008**, *26*, 147–162. [CrossRef]

35. USGS. *Commodity Statistic and Information*; USGS: Reston, VA, USA, 2018.

36. D'Adamo, I.; Miliacca, M.; Rosa, P. Economic feasibility for recycling of waste crystalline silicon photovoltaic modules. *Int. J. Photoenergy* **2017**, *2017*, 1–7. [CrossRef]

37. Li, Z.; Zhao, Y.; Wang, X.; Sun, Y.; Zhao, Z.; Li, Y.; Zhou, H.; Chen, Q. Cost Analysis of perovskite tandem photovoltaics. *Joule* **2018**, *2*, 1559–1572. [CrossRef]

© 2020 by the authors. Licensee MDPI, Basel, Switzerland. This article is an open access article distributed under the terms and conditions of the Creative Commons Attribution (CC BY) license (http://creativecommons.org/licenses/by/4.0/).

Article

Life Cycle Assessment of Italian Electricity Scenarios to 2030

Alessia Gargiulo *, Maria Leonor Carvalho * and Pierpaolo Girardi *

Ricerca Sistema Energetico–RSE SpA, 20134 Milan, Italy
* Correspondence: alessia.gargiulo@rse-web.it (A.G.); marialeonor.carvalho@rse-web.it (M.L.C.); pierpaolo.girardi@rse-web.it (P.G.)

Received: 26 June 2020; Accepted: 24 July 2020; Published: 28 July 2020

Abstract: The study presents a Life Cycle Assessment (LCA) of Italian electricity scenarios, devised in the Integrated National Energy and Climate Plan (INECP). A fully representative LCA of the national electricity system was carried out, taking into consideration a great number of different power plant typologies for current (2016 and 2017) and future (2030) electricity mixes. The study confirms that LCA can be a powerful tool for supporting energy planning and strategies assessment. Indeed the results put in evidence not only the improvement of the environmental profile from the current to the future mix (the impacts decrease from 2016 to 2030 due to the transition towards renewables, mainly wind and photovoltaic), but also underline the difference between two scenarios at 2030 (being the scenario that includes the strategic objectives of the INECP to 2030 the one showing best environmental profile), providing an evaluation of the effect of different energy policies. For example, in the INECP scenario CO_2 eq/kWh is 46% lower than current scenario and 37% lower than business as usual scenario for 2030. Moreover, considering different impact categories allowed to identify potential environmental trade-offs. The results suggest also the need of future insight on data related to photovoltaic technologies and materials and their future development.

Keywords: electricity scenarios; life cycle assessment; Italian electricity; environmental impacts

1. Introduction

European Commission energy policy at the 2030 horizon aims to strengthen the 20-20-20 objectives and, at the same time, is a precondition for 2050 goals of the long-term strategy to reduce greenhouse gas emissions [1]. In this framework, the Italian Integrated National Energy and Climate Plan (INECP) [2], intends to accelerate the transition from traditional fuels to renewable sources.

Due care must be paid to ensure that the energy and climate objectives are compatible with the objectives relating to the landscape protection, the quality of air and water bodies, the safeguarding of biodiversity and soil protection. As a matter of fact, the necessary measures to increase decarbonisation of the system, involve power plants and infrastructure deployment that have environmental impacts [2].

The present study is aimed at evaluating, from an environmental point of view, the Italian electricity generation scenarios at 2030 (devised in the INECP) and at comparing them with the current electricity generation mix.

At this purpose, Life Cycle Assessment (LCA) methodology according to ISO 14040 [3] has been adopted. According to ISO 14040 [3] LCA is a methodology that *"addresses the environmental aspects and potential environmental impacts throughout a product's life cycle from raw material acquisition through production, use, end-of-life treatment, recycling and final disposal (i.e., cradle-to-grave)"*. LCA can be a powerful tool for supporting energy planning for several reasons:

- For an effective energy transition towards a low carbon system, the reduction of greenhouse gas emissions shall occur over the entire chain of energy production and consumption. LCA allows to evaluate the potential impacts of a system/product taking into account all the processes along the entire life cycle.

- LCA provides an impact assessment of the electricity scenarios not only on climate change, but taking into account a more complex environmental profile, including several environmental impact categories, in order to verify the effects of a policy on the different environmental aspects and to point out potential environmental trade-offs [4].
- An in-depth LCA analysis of the current and future electricity mix is definitely relevant since energy policies promote the electrification of final energy consumptions. The detailed and updated analysis of the current and future mix can be used as a reference in other LCA studies, since the electricity production pervades the life cycle of numerous products and often represents one of the most relevant processes.
- LCA results (detailed for phase and geographic location) represent the basis for monetary valuation of environmental externalities [5].

For these reasons, LCA have been widely used in literature to assess present and future national electricity mix scenarios sustainability [6–9]. Some works put also in evidence issues and explore new methodological solutions. For example, the study [10] on the life-cycle assessment of the large-scale implementation of climate-mitigation technologies, addresses the impacts on the electricity and uses assumptions of technical improvements also in material production technologies. Reference [11] combines different approaches in a "technology hybridized environmental-economic model with integrated scenarios", to predict the environmental impacts of energy policy scenarios. Recent studies evaluate, with a life cycle approach, energy scenarios at a national (Spain [6] and Germany [7]) or regional (Sicilia region in Italy [9]) scale. According to [6], that provides an investigation into the sustainability of the electrical system in Spain, for future scenarios (2030 and 2050), the most ambitious projections in terms of renewable penetration perform best in terms of environmental performance and the scenario considering higher fossil fuel contributions performs worst in all sustainability indicators. As demonstrated in [12] the 2030 New York scenario, based on 70% of renewable energies, dramatically reduces both carbon dioxide emissions and cumulative energy demand. Finally, the life cycle assessment of UK electricity scenarios to 2070 was studied by [8]. According to the LCA results, the decarbonisation of the UK electricity mix introduces many questions regarding sustainability and shows that the level of decarbonisation achieved and the method taken can lead to significantly diverging outcomes, each involving trade-offs and compromises.

In this framework, the present study evaluates the Italian energy strategy, starting from a detailed and fully representative LCA of the Italian electricity system. In order to represent the variability of energy sources, fuels and transformation technologies, the study takes into consideration a great number of different electricity power plant types for current and future electricity mixes. For operation phase of fossil thermoelectric sector, updated primary data for the main air emissions have been used. In fact, also considering the increasing role of electricity as an energy carrier, data quality and representativeness, is a crucial issue in life cycle inventory of electricity supply [13].

Two scenarios at 2030 have been used in this work, in order to evaluate the effect of the electricity system evolution on the environmental indicators. These scenarios are described and utilized as the basis of the Italian Integrated National Energy and Climate Plan [2]. For Italian current electricity system, two years are taken into consideration: 2016 and 2017. Year 2016 is the base year used for INECP scenarios elaboration, while year 2017 is the most recent year for which statistical data were available (when performing the study) and it is considered in order to present an updated LCA of present Italian electricity system.

Goal and scope of the LCA, inventory and impact assessment results are described in the following paragraphs.

2. Materials and Methods

The goal of the present study is the Life Cycle Assessment (ISO 14040 [3]) of electricity generation scenarios in Italy at 2030. In particular, two scenarios developed for the INECP are taken into consideration: the Baseline scenario (2030 BASE) that describes an evolution of the Italian energy system with current policies and measures and the INECP scenario (2030 INECP) that quantifies the strategic objectives of the plan [2].

In this study each scenario is defined by:

-   A mix of electricity production technologies and energy sources;
-   The efficiency and load factor related to the specific technology.

For future mixes, the technological progress in the electricity conversion technologies was taken into account through enhanced conversion efficiencies and load factors.

Background system evolution in time (as for example global production market of main materials for power plant construction) has not been considered in this study. This can be a critical issue especially for studies on long term scenarios [14]. Nevertheless the results of the present study could help to identify the most relevant background processes and materials production markets, with a role in environmental potential trade-offs which can be investigated in future studies, with a much longer perspective (2050, 2070).

In order to evaluate the evolution of the electricity mix, and consequently the evolution of the environmental profile associated with it, the LCA of the current electricity mix is first presented, taking into consideration years 2016 and 2017.

The functional unit is 1 kWh of electricity Gross National Consumption (GNC) which includes the total gross national electricity generation from all sources (excepted pumped hydro generation), plus electricity imports, minus exports. As regards system boundaries, all phases of the life cycle, from cradle to grave, are included in the analysis: fuel supply, power plant construction, power plant operation and power plant end-of-life.

Since the functional unit refers to GNC and not to the final consumption, transmission and distribution network, as well as the losses associated with it, are excluded from the boundaries of the analyzed system. The impact categories and assessment methods are selected on the basis of Impact Assessment guidelines [15] drawn up by Joint Research Center—European Commission—JRC. Only impact categories reported in the guideline with the level of recommendations I (recommended and satisfactory) and II (recommended but in need of some improvements) are utilized (see Table 1). For assessment methods description and reference, refer to [15].

Table 1. Impact categories taken into consideration in the LCA of the Italian electricity scenarios.

Impact Category	u.m.	Assessment Method	Recommendation Level
Climate Change	(kg CO_2 eq)	Baseline model of 100 years of the IPCC	I
Ozone depletion	(kg CFC^{-11} eq)	Steady-state ODPs	I
Particulate matter	(kg $PM_{2.5}$ eq)	RiskPoll model	I
Ionizing radiation HH	(kBq U235 eq)	Human health effect model	II
Photochemical ozone formation	(kg NMVOC eq)	LOTOS-EUROS as applied in ReCiPe	II
Acidification	(molc H^+ eq)	Accumulated Exceedance	II
Terrestrial eutrophication	(molc N eq)	Accumulated Exceedance	II
Freshwater eutrophication	kg P eq	EUTREND model as implemented in ReCiPe	II
Marine eutrophication	kg N eq	EUTREND model as implemented in ReCiPe	II
Mineral, fossil & ren resource depletion	kg Sb eq	CML 2002	II

Regarding data quality, primary data (statistical data and environmental declarations from Italian power plants) and secondary data (Ecoinvent LCI database [16]) were used, as better specified below. For elaborations the LCA software SimaPro (v8, PRè Consultant, Amersfoort, The Netherlands) was used.

As regards the allocation of impacts between the main product and by-products, the "cut-off" [16] approach was adopted. To the secondary (recycled) materials, only the impacts of the recycling process are assigned (no impact from the primary production of the material). In the case of electricity from wastes, all the impacts of incineration are allocated to the waste treated in the plant (electricity production is burden free). For allocation between heat and electricity in the cogeneration power plants, allocation proportional to the output was used. The fuel is attributed proportionally to the amount of the output product (electricity and heat). This approach was chosen since it is coherent to the method used in Eurostat energy balance.

Statistical data have been elaborated in order to obtain electricity mix detailed by power plant typology. A power plant typology is defined by a combination of fuel in input and transformation technology (e.g., natural gas combined cycle power plant). Since available official data and life cycle inventories present aggregated data for Italian electricity mix, a combination of different official energy statistics has been analyzed and elaborated, in order to consider a complete set of fuels and technologies. As regards thermal power plants, technologies taken into consideration are those reported in the statistical reports published annually by TERNA (the Italian system transmission operator) [17]:

- electricity production only (Only EL): internal combustion (CI), gas turbine (TG), condensing steam (C), combined cycle (CC), repowered (RP);
- combined heat and power production (CHP): internal combustion (CIC), gas turbine (TGC), combined cycle (CCC), counter pressure steam (CPC), condensing steam with bleed (CSC).

Fuels taken into consideration are all those reported in the energy balance published by Eurostat [18]. In addition to thermoelectric (fossil and renewable), the mix includes hydroelectric (reservoir and runoff), wind and photovoltaic plants.

Tables A1–A4, in Appendix A, show Italian current and future electricity mixes, with the details of the power plant typology. The 2016 and 2017 mixes are very similar: natural gas accounted for 39–42% of the total electricity production in Italy. Among renewable sources, hydropower ranks first, covering 11–13% of the total production, followed by solar and wind energy. A share of about 11% of the electricity is imported. For scenario BASE at 2030 a greater penetration of renewable is foreseen especially for hydropower (from 10% to more than 15%), wind power (from 5% to 7.5%) and solar power (from 7% to almost 10%). In the INECP Scenario a phase out of coal is included, with zero contribution at 2030, while the penetration of solar power rises up to more than 20%.

To build the life cycle inventory, both primary and secondary data have been taken into account. For future mixes, the technological progress in the electricity conversion technologies was taken into account through conversion efficiencies and load factors, resulting from the scenarios described in the INECP [2]. Table 2 contains power plant efficiencies taken into consideration for current and future mixes.

Primary data include:

- electricity production and efficiency for each type of power plant (Table 2): Eurostat [18] and TERNA [17] statistical data were used for current mix; scenarios data from the INECP [2] were used for 2030 mixes;
- wind farm load factor: Eurostat data [18] were used for current mix; scenario data from the INECP [2] were used for 2030 mixes (Table 3);
- photovoltaic load factor: Eurostat data [18] were used for current mix; scenario data from the INECP were used for 2030 mixes (Table 3);
- natural gas import market: SNAM data (Sustainability Report and Ten-Year Development Plan) [19] were used;
- oil products input market: data from the Italian Oil Association [20] were used;
- CO_2 emissions during operation of the thermoelectric plants: ISPRA [21] values based on IPCC [22] were used with the exception of "Other Petroleum power plants", for which we used

The values based on environmental declarations of the Italian thermoelectric plants registered to the Community eco-management and audit system—EMAS (Regulation 1221/2009) [23]. Average emission factors per unit of fuel in input (used in this study for current and future scenarios) are reported in Table 4.

Table 2. Power plant electrical efficiencies taken into consideration for current and future mixes.

Fuel	Scenario	CI	TG	C	CC	CIC	TGC	CCC	CPC	CSC
Other bituminous coal	2016			0.39				0.25		
	2017			0.39				0.25	0.11	
	2030 BASE			0.47						
Sub bituminous coal	2016			0.39						
	2017			0.38						
Coke oven gas	2016					0.35		0.32		0.34
	2017					0.38		0.33		0.36
Blast furnace gas	2016					0.32		0.29		0.32
	2017					0.34		0.30		0.32
	2030 BASE							0.30		
	2030 INECP							0.30		
Other recovered gas	2016					0.33		0.30		0.33
	2017					0.36		0.31		0.33
Refinery gas	2016					0.39	0.30	0.26	0.12	0.12
	2017					0.35	0.28	0.26	0.10	0.14
Liquefied petroleum gases	2016							0.31		
	2017						0.35	0.32		
Gas oil and diesel oil (without biofuels)	2016	0.39		0.36				0.47		
	2017	0.39		0.36	0.42			0.47		
Fuel oil	2016	0.38		0.35			0.18	0.16		
	2017	0.38		0.35			0.18	0.16		0.10
Other oil products	2016			0.24				0.23		
	2017			0.23				0.23		
	2030 BASE							0.48		
	2030 INECP							0.49		
Natural gas	2016	0.37	0.32	0.38	0.54	0.41	0.32	0.48	0.20	0.27
	2017	0.37	0.31	0.39	0.54	0.41	0.32	0.49	0.20	0.28
	2030 BASE				0.58	0.45	0.36	0.50		
	2030 INECP				0.54	0.45	0.36	0.50		
Primary solid biofuels	2016	0.41		0.26		0.21			0.20	0.13
	2017	0.42		0.28		0.21			0.20	0.14
	2030 BASE			0.33		0.39				
	2030 INECP			0.33		0.39				
Biogas	2016	0.37	0.32			0.40		0.36		
	2017	0.37	0.32			0.40		0.36		0.16
	2030 BASE	0.40	0.40			0.40				
	2034 INECP	0.40	0.40			0.40				
Renewable municipal waste	2016			0.25		0.31			0.30	0.20
	2017			0.25		0.30			0.29	0.20
	2030 BASE					0.23			0.23	0.23
	2030 INECP					0.23			0.23	0.23
Other liquid biofuels	2016	0.41			0.49	0.40			0.40	0.36
	2017	0.42			0.48	0.40			0.40	0.37
Industrial wastes	2016			0.26						0.14
	2017			0.26		0.22	0.10			0.14
Non-renewable municipal waste	2016			0.25		0.31			0.30	0.20
	2017			0.25		0.30			0.29	0.20
	2030 BASE					0.23			0.23	0.23
	2030 INECP					0.23			0.23	0.23

Table 3. Wind and Photovoltaic load factors for current and future mixes.

Energy Source	2016	2017	2030 Base	2030 INECP
Wind	1885	1822	2029	2029
Photovoltaic	1146	1239	1329	1329

Table 4. CO_2 average emission factors per unit of fuel in input, in operation phase of power plants.

Power Plant Type	CO_2 (kg/MJ$_{in}$)
Natural Gas	0.0564
Coal	0.0939
Fuel Oil	0.0767
Gas Diesel Oil	0.0741
LPG	0.0642
Pet. TAR	0.1211

-   NOx, SOx and PM10 emissions during operation of thermoelectric fossil power plants: a selection of data extracted from the Environmental Declaration (requested by EMAS Regulation) of Italian Thermoelectric plants were elaborated. The elaboration included 102 plants, which constitutes a sample of the Italian thermoelectric power plants. About half of the power plants were excluded from the calculation of the average emissions due to incompleteness or because they are multi-fuel plants. The sample of power plants used for the calculation of the average emissions covers from 30% (oil power plants) to about 90% (coal power plants) of the total electricity production in Italy (year 2017), depending on the type of power plants, as reported in Table 5. Average emission values are reported in Table 6.

Table 5. Description of the sample of power plants used for the calculation of the average emissions.

Power Plant Type	N° Power Plants in the Sample	Electricity Production (% on Total) [1]
Natural Gas	35	53%
Coal	8	93%
Fuel Oil	1	30%
Other Oil products	1	56%
Total (fossil thermoelectric)	**45**	**56%**

[1] Sum electricity production of the sample/total electricity production.

Table 6. Calculated average emissions per unit of fuel in input, in operation phase of power plants.

Power Plant Type	SOx (g/MJ$_{in}$)	NOx (g/MJ$_{in}$)	PM 10 (g/MJ$_{in}$)
Natural Gas	0.00	1.73×10^{-2}	1.88×10^{-5}
Coal	2.99×10^{-2}	3.99×10^{-2}	1.25×10^{-3}
Fuel Oil	3.22×10^{-2}	2.47×10^{-2}	3.33×10^{-3}
Pet TAR	1.51×10^{-2}	2.06×10^{-2}	2.03×10^{-4}

Exceptions are coke oven gas and blast furnaces gas for which secondary (Ecoinvent 3.3 [16]) data have been used.

Secondary data from database Ecoinvent [16] have been used for power plants construction and dismantling, as well as for all background systems.

3. Results

The effects of policies in the INECP scenario are very evident compared to the baseline scenario. Both 2030 scenarios lead to an increase in renewables, but the strategic objectives of the plan in INECP scenario make the transition to renewables decidedly evident, bringing to zero the electricity production from coal and driving wind plus photovoltaic share in the mix to more than 30%.

In the following graph (Figure 1) the percentage contributions of the different power plants to the mix are highlighted and the electricity mix by the European Commission scenario PRIMES 2016 (2030 EU-REF IT) [24] is also added for comparison. The EU Reference Scenario is one of the European

Commission's key analysis tools in the areas of energy, transport and climate action. It uses the PRIMES model for energy and CO_2 projections [24].

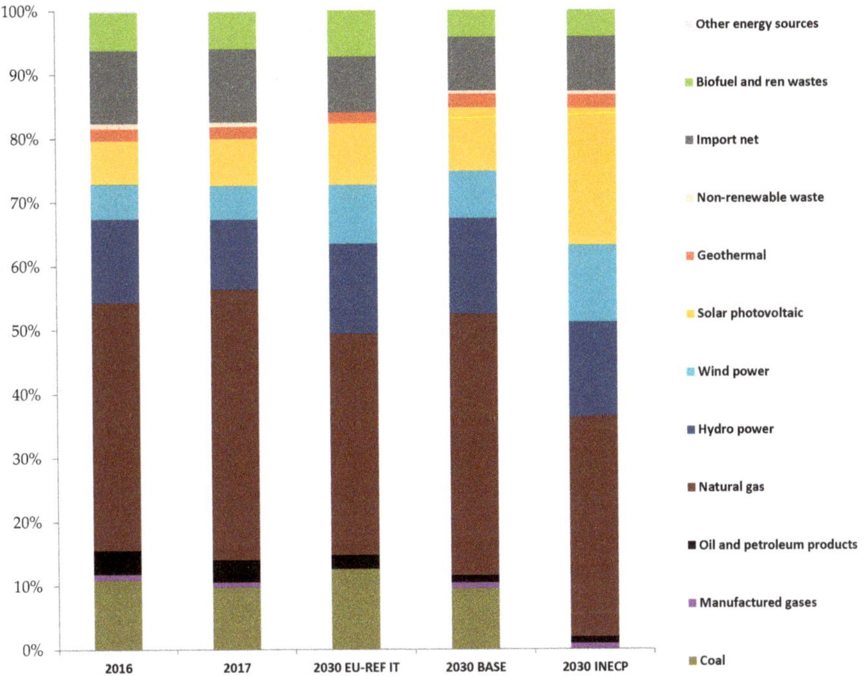

Figure 1. Contribution of different power plants to current and future electricity mixes.

Table 7 shows the results of the impact assessment. For each category, the impact along the entire life cycle of the electricity mix is reported, referred to the functional unit, 1 kWh of GNC.

Table 7. Life Cycle Impact Assessment results per 1 kWh of electricity GNC.

Impact Category	u.m.	2016	2017	2030 BASE	2030 INECP	2030 EU-REF IT
Climate Change	kg CO_2 eq	4.18×10^{-1}	4.17×10^{-1}	3.57×10^{-1}	2.26×10^{-1}	3.14×10^{-1}
Ozone depletion	kg CFC-11 eq	3.79×10^{-8}	4.05×10^{-8}	3.72×10^{-8}	3.16×10^{-8}	3.17×10^{-8}
Particulate matter	kg PM2.5 eq	7.47×10^{-5}	7.48×10^{-5}	6.30×10^{-5}	5.73×10^{-5}	6.10×10^{-5}
Ionizing radiation HH	kBq U235 eq	2.87×10^{-2}	2.89×10^{-2}	1.98×10^{-2}	1.96×10^{-2}	2.02×10^{-2}
Photochemical ozone formation	kg NMVOC eq	4.67×10^{-4}	4.71×10^{-4}	4.08×10^{-4}	3.14×10^{-4}	3.80×10^{-4}
Acidification	molc H$^+$ eq	8.97×10^{-4}	8.93×10^{-4}	7.48×10^{-4}	5.69×10^{-4}	7.23×10^{-4}
Terrestrial eutrophication	molc N eq	1.63×10^{-3}	1.61×10^{-3}	1.41×10^{-3}	1.06×10^{-3}	1.39×10^{-3}
Freshwater eutrophication	kg P eq	9.41×10^{-5}	9.07×10^{-5}	7.43×10^{-5}	4.55×10^{-5}	7.80×10^{-5}
Marine eutrophication	kg N eq	1.88×10^{-4}	1.84×10^{-4}	1.29×10^{-4}	9.51×10^{-5}	1.25×10^{-4}
Mineral, fossil & ren resource depletion	kg Sb eq	3.22×10^{-6}	3.18×10^{-6}	3.33×10^{-6}	6.05×10^{-6}	3.05×10^{-6}

For the sake of completeness, a comparison is also provided with the 2030 scenario developed by the European Commission for Italy (2030 EU-REF IT).

The INECP scenario is the one with the best environmental performance, resulting in the least impact for almost all categories. The only notable exception is the impact category "Mineral, fossil & ren resource depletion": the strong increase in photovoltaic (more than double in percentage compared to the baseline scenario) is the main reason for the greater impact and is essentially due to the metals present in the inverter and to the aluminium frame and the support structures of the modules. This impact could be significantly reduced in the future thanks to the diffusion of innovative photovoltaic solutions, as for example double-sided glass-glass modules. It should be reminded that in the present study only the evolution of power plant efficiencies and load factors is taken into consideration in the scenarios, and no hypothesis is formulated about changes in the background system, as for example global production market of main materials such as aluminium. This could be an interesting task for future insights, above all on wind and photovoltaic technologies, for which background processes have a higher impact than operation and maintenance phase.

Results (Figure 2) show a general decrease from 2016 to 2030 of the impacts of the Italian electricity mix. The most marked decrease is observed for Climate Change (−46% compared to 2016 in the INECP case) and Water Eutrophication (−51% compared to 2016 in the PNIEC case) impact categories. The decrease is driven by the transition to renewables (mainly wind and photovoltaic).

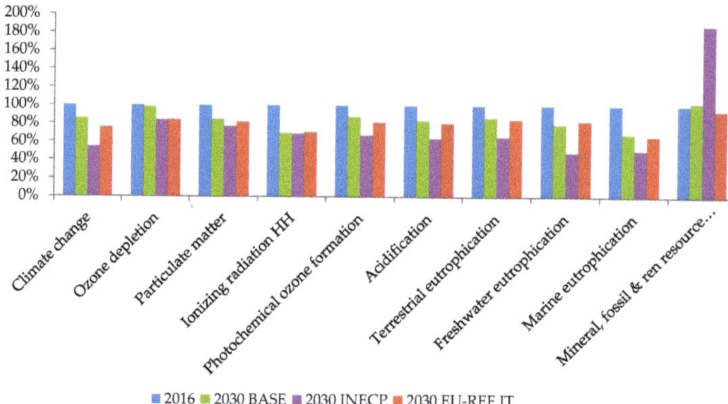

Figure 2. Comparison between Life cycle impact assessment of current and future mixes. Results in percentage respect to 2016.

The decrease of Ionizing Radiation impact is due to the lowest share of imported electricity, since for this impact category, nuclear energy (absent in the Italian mix but present in the European import mix) has the greatest effect.

For all the impact categories, BASE scenario results are similar to the results coming from EU-REF IT scenario.

Table 8 highlights the category of power plants that most contributes to the impact, for both scenarios at 2030.

Table 8. Main contributor to the impact for Base and INECP scenarios.

Impact Category	Main Contributor to the Impact
Climate Change	Natural gas power plants
Ozone Depletion	Natural gas power plants
Particulate Matter	Natural gas power plants and Photovoltaic (INECP)
Ionizing radiation	Imported electricity
Photochemical Ozone Formation	Natural gas power plants
Acidification	Natural gas power plants
Terrestrial eutrophication	Natural gas power plants
Marine eutrophication	Natural gas power plants
Freshwater eutrophication	Coal power plants(2030 BASE) Net Import (INECP)
Mineral fossil and renewable resource depletion	Photovoltaic system

Each technology contribution is determined by two factors: the specific impact of the single source/production technology and the share of the single source/production technology in the electricity mix. Natural gas power plants contribution is in general due to the high share in the mixes.

More in detail, for both scenarios at 2030, BASE and INECP:

- Climate change: main contribution comes from electricity produced by natural gas power plants.
- Ozone depletion: main contribution comes from electricity produced by natural gas power plants (it is mainly related to fuel upstream phase, and in particular to the transport of gas through pipeline and to the refrigerants used in the compression stations).
- Particulate matter: main contribution comes from electricity produced by natural gas power plants; in the INECP scenario it comes also from photovoltaic, due to the market for silicon wafer production. Emissions are linked to China energy mix, used for photovoltaic silicon wafer production.
- Ionizing radiation: as above mentioned, main contribution comes from imported electricity.
- Photochemical ozone formation: main contribution comes from electricity produced by natural gas power plants.
- Acidification: in addition to natural gas power plants also the impacts relating to imported electricity and coal plants (for 2030 BASE only) should be highlighted. The impacts are substantially associated with the coal upstream (also in the case of imported electricity).
- Terrestrial eutrophication and marine eutrophication: main contribution is associated to natural gas, coal power plants and import. It must be underlined also the contribution deriving from biomass power plants (this is due in part to the cultivation phase of vegetables dedicated to bioliquid production).
- Freshwater eutrophication: main contribution comes from coal upstream. According to background life cycle inventory data, the impact is mainly due to coal mining operations. As concerns INECP main contribution comes from Net Import.
- Mineral fossil and renewable resource depletion: main contribution comes from photovoltaic systems, due to the aluminium used for the frame and for the support structure of the modules, as well as to the metals present in the inverter.

Figures 3 and 4 put in evidence, for each impact category, the relative contribution of each energy source.

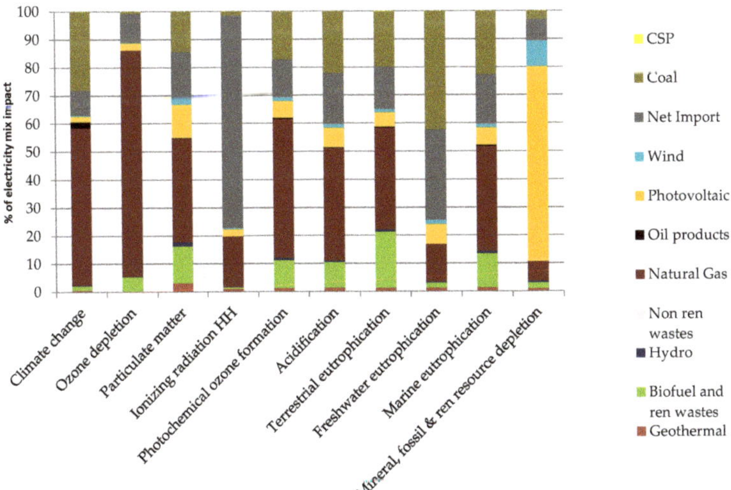

Figure 3. Contribution of different typologies of power plant to the overall impact of the 2030 BASE scenario electricity mix for Italy.

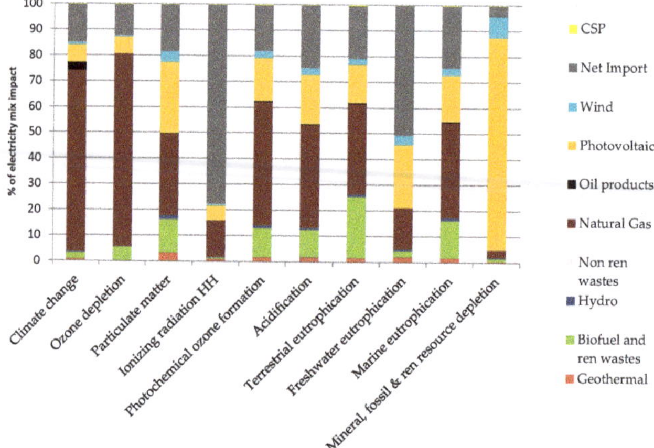

Figure 4. Contribution of different typologies of power plant to the overall impact of the 2030 INECP scenario electricity mix for Italy.

More in detail, regarding climate change, annual CO_2 eq emissions due to Italian electricity mix amount to 136–138 Mt/year in 2016–2017, 122 Mt/year in 2030 for the BASE scenario, and 76 Mt/year in 2030 for the INECP scenario.

The main driver of this trend is the decrease in the share of electricity produced from coal and, to a lesser extent, from oil and imported electricity. Finally the increase in the average efficiency of the generation mix also plays a role in reducing CO_2 eq emissions.

The contribution of natural gas power plants to the CO_2 eq emissions is strongly influenced by the fact that the share in the mix of electricity produced by natural gas power plant stands at high values (from 34% to 42%). On the other hand, coal plants with a much lower share in the mix, 9–12% (in the case of the INECP scenario it is equal to 0%), cover from 25% to 35% of the impact. In no case do wind, photovoltaic and hydropower contribute significantly to the CO_2 eq emissions of the mix, despite of shares in the mix far from negligible (10–15% for hydroelectric, 5–11% for wind, 7–21% for photovoltaic) (Figure 5).

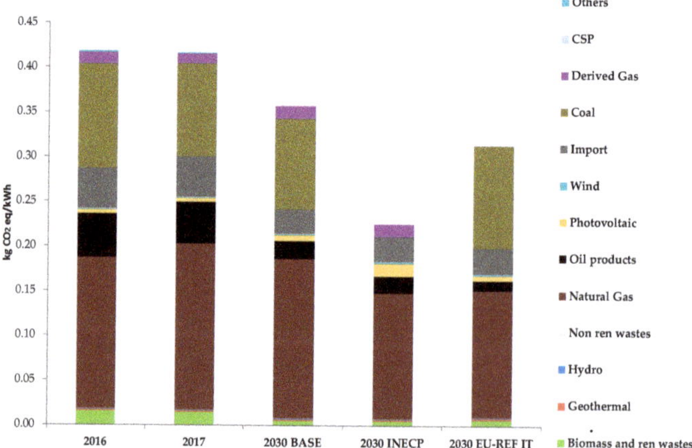

Figure 5. CO_2 eq/kWh for current and future electricity mixes. The contribution of different typologies of power plant is highlighted.

4. Discussion and Conclusions

According to ISO 14040, LCA results can be elaborated trough two optional steps: normalization and weighing. Both can help in interpretation of the results, but can add elements of subjectivity to the evaluation.

In this study, it was decided to leave out weighting, which should involve decision-makers in the process. Conversely, normalization was carried out, which allows to rank and compare the different impacts of a system and, although not devoid of limitation, is a great aid in looking at scenarios environmental profile as a whole.

Even though the limitations reported in [25] have to be considered, normalisation has a relevant role when LCA is aimed at supporting policy makers to ensure that the focus is put on most relevant aspects and for communication purposes.

In normalization JRC's recommendations [25] have been taken into consideration. In view of the international nature of energy supply chains, it was decided to use normalization factors on a global scale. The normalisation factors represent the total impact of a reference region for a certain impact category (e.g., Climate Change, Eutrophication, etc.) in a reference year.

The standardization factors used are those implemented in the Simapro v.8 software and referring to the JRC table version 0.1.1-15/12/2015 (available from: https://eplca.jrc.ec.europa.eu/LCDN/developerILCD. xhtml), developed as the first version in the context of the work described in [25].

For each impact indicator, the impact assessment result is divided by the global value (i.e., deriving from all human activities) of the same indicator, on a per capita basis. In order to apply the normalization, the Italian gross national electricity consumption per capita per year (electricity yearly GNC divided by the Italian population) was calculated for current and future scenarios. The following graphs show normalized LCA results (dimensionless).

In all scenarios (current, i.e., 2016 and 2017 results are very similar. In the graph only 2016 results are shown, 2030 BASE and 2030 INECP), most of the impact categories present similar values (lower than 0.10). It is notable that, also in the case of water eutrophication impact categories for which life cycle impact assessment showed the higher reduction from current to INECP scenario, the normalized results are lower than 0.10. On the other hand the Climate Change category and the Ionizing Radiation category present definitely higher values (0.18–0.33 and 0.45–0.66 respectively).

The impact on resource depletion, the only category which increase along the time horizon of the assessment, remains, as normalized value, under 0.20, also in the case of INECP scenario (Figures 6–8)

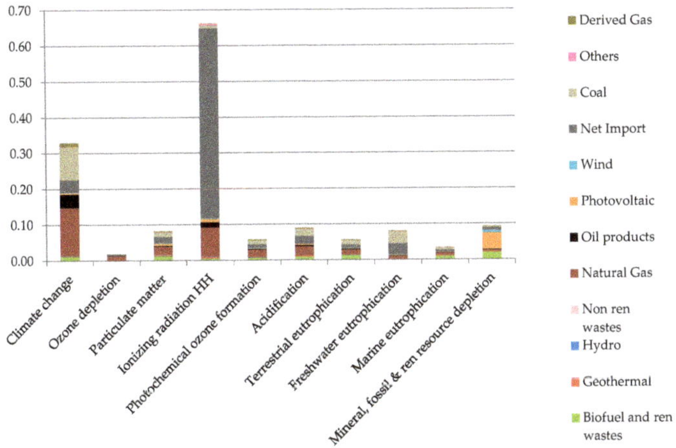

Figure 6. Normalized results for current Italian gross national (2016) electricity consumption per capita per year.

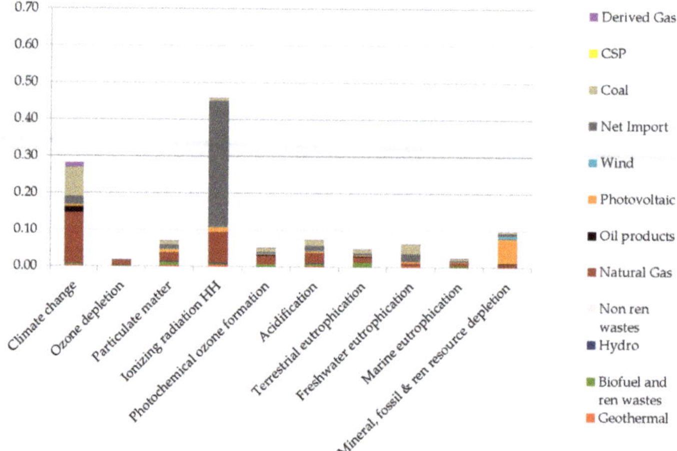

Figure 7. Normalized results for 2030 BASE Italian gross national electricity consumption per capita per year.

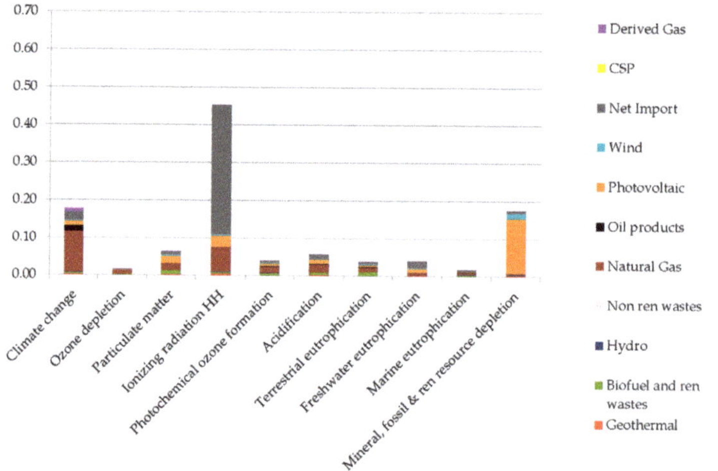

Figure 8. Normalized results for 2030 INECP Italian gross national electricity consumption per capita per year.

In conclusion, the present study confirms that LCA can be a powerful tool for supporting energy planning and strategies assessment. As a matter of fact, results put in evidence not only the improvement of the environmental profile from the current to the future mix, but also underline the difference between the baseline scenario and the INECP one, providing an evaluation of the effect of different energy policies. According to LCA results, the impacts of the Italian electricity mix decrease from 2016 to 2030 due to the transition to renewables, mainly wind and photovoltaic. Climate Change impact decreases by about 46% compared to 2016. Most important for policy implication, the scenario that includes the strategic objectives of the Integrated National Plan for Energy and Climate to 2030 is the one with the best environmental profile (INECP scenario CO_2 eq/kWh are 37% lower than 2030 BASE scenario). Moreover, considering not only climate change but a set of different impact categories, allowed to identify potential environmental trade-off, thus obtaining a more complete environmental assessment of energy policies.

The only potential environmental trade-off (even if slight looking at normalized results), seems to occur between climate change and the impact category related to resources depletion. The impact on resource depletion is mainly associated with the metals present in the inverter and especially with the aluminium frame and structure of the photovoltaic modules. This finding can address subsequent studies and insights. First of all, a big effort is demanded for improving and updating the inventory data relating to the photovoltaic modules (in the present study, only secondary data were used for the construction and end-of-life phase). Particular attention must also be paid to the aspects relating to the recycling processes (especially for aluminum) also from a methodological point of view (allocation of impact on primary or secondary materials), since a significant reduction in resource depletion impact may depend on recycling. Photovoltaic only in relatively recent years deeply penetrates energy mixes and for this reason, data on recycling and on the secondary products market are not widely available. Sensitivity analysis on various recycling hypotheses could be useful [26].

Finally, in the context of improving inventory data, the technological evolution towards new photovoltaic solutions (for example heterojunction modules [27]) should also be taken into consideration especially in the case of longer-term scenarios, like those in 2050, a horizon identified by Italy for a deep decarbonisation of the energy sector. As mentioned, beside assessing the environmental effect of the energy policy, the study demonstrates that LCA is a powerful tool in supporting decision makers especially dealing with future national energy plan. Nevertheless, although well beyond the scope of the present study, it should be underlined that also economic and social impacts must be taken into account in policy and decision making in energy sector [6]. Sustainable development of energy systems, in fact, requires that all three pillars of, environmental, economic and social, are taken into consideration [28].

Author Contributions: Conceptualization, A.G. and P.G.; methodology, A.G.; validation, A.G. and P.G.; formal analysis, A.G. and M.L.C.; investigation, A.G.; data curation, A.G.; writing—original draft preparation, A.G.; writing—review and editing, A.G., P.G. and M.L.C.; visualization M.L.C.; supervision, P.G.; project administration, P.G.; All authors have read and agreed to the published version of the manuscript.

Funding: This work has been financed by the Research Fund for the Italian Electrical System in compliance with the Decree of Minister of Economical Development 16 April 2018.

Acknowledgments: This work has been financed by the Research Fund for the Italian Electrical System in compliance with the Decree of Minister of Economical Development 16 April 2018.

Conflicts of Interest: The authors declare no conflict of interest.

Appendix A

Table A1. Current (2016) Italian electricity mix.

Source	Total %	Only EL %	CI %	TG %	C %	CC %	CHP %	CIC %	TGC %	CCC %	CPC %	CSC %
Coal	10.96	10.91	-	-	10.91	-	0.04	-	-	0.04	0.00	-
Other bituminous coal	10.88	10.84	-	-	10.84	-	0.04	-	-	0.04	0.00	-
Sub-bituminous coal	0.08	0.08	-	-	0.08	-	-	-	-	-	-	-
Manufactured gases	0.86	-	-	-	-	-	0.86	0.04	-	0.43	-	0.39
Coke oven gas	0.23	-	-	-	-	-	0.23	0.01	-	0.12	-	0.11
Blast furnace gas	0.62	-	-	-	-	-	0.62	0.03	-	0.31	-	0.29
Oil and petroleum products	3.73	0.62	0.08	0.00	0.54	0.00	3.11	0.01	0.13	2.89	0.04	0.05
Refinery gas	0.61	-	-	-	-	-	0.61	0.01	0.11	0.41	0.03	0.04
Liquefied petroleum gases	0.01	-	-	-	-	-	0.01	-	0.00	0.01	-	0.00
Gas oil and diesel oil (without biofuels)	0.13	0.12	0.02	-	0.11	-	0.01	-	0.00	0.01	0.00	0.00
Fuel oil	0.52	0.47	0.06	0.00	0.40	0.00	0.06	0.00	0.01	0.04	0.00	0.00
Other oil products	2.46	0.03	-	-	0.03	-	2.43	-	-	2.43	-	-
Natural gas	38.82	13.84	0.08	0.11	0.11	13.55	24.97	2.50	1.37	20.66	0.23	0.21
Renewables and biofuels	33.24											
Hydro power	13.06											
Wind power	5.44											
Solar photovoltaic	6.80											
Geothermal	1.94											
Primary solid biofuels	1.27	0.69	0.09	-	0.60	-	0.58	0.17	0.00	-	0.06	0.35
Biogas	2.54	0.95	0.93	0.01	0.00	0.00	1.60	1.59	0.00	0.01	-	0.00
Renewable municipal waste	0.74	0.38	-	-	0.38	-	0.37	0.11	0.00	-	0.04	0.22

Table A1. *Cont.*

Source	Total %	Only EL %	CI %	TG %	C %	CC %	CHP %	CIC %	TGC %	CCC %	CPC %	CSC %
Other liquid biofuels	1.45	1.01	0.34	0.00	-	0.67	0.43	0.13	0.00	-	0.05	0.26
Non-renewable waste	0.77	0.39	-	-	0.39	-	0.38	0.11	0.00	-	0.04	0.23
Industrial wastes	0.03	0.02	-	-	0.02	-	0.01	0.00	-	-	0.00	0.01
Non-renewable municipal waste	0.74	0.38	-	-	0.38	-	0.37	0.11	0.00	-	0.04	0.22
Import net	11.40											
other energy sources	0.23											
Electricity GNC	100.00											

Table A2. Current (2017) Italian electricity mix.

Source	Total %	Only EL %	CI %	TG %	C %	CC %	CHP %	CIC %	TGC %	CCC %	CPC %	CSC %
Coal	9.84	9.79	-	-	9.79	-	0.04	-	-	0.04	0.00	-
Other bituminous coal	9.83	9.79	-	-	9.79	-	0.04	-	-	0.04	0.00	-
Sub-bituminous coal	0.01	0.01	-	-	0.01	-	-	-	-	-	-	-
Manufactured gases	0.74	-	-	-	-	-	0.74	0.04	-	0.40	-	0.30
Coke oven gas	0.25	-	-	-	-	-	0.25	0.01	-	0.14	-	0.10
Blast furnace gas	0.44	-	-	-	-	-	0.44	0.02	-	0.24	-	0.18
Other recovered gas	0.05	-	-	-	-	-	0.05	0.00	-	0.03	-	0.02
Oil and petroleum products	3.47	0.61	0.07	0.00	0.54	0.00	2.87	0.01	0.13	2.62	0.03	0.07
Refinery gas	0.57	-	-	-	-	-	0.57	0.01	0.11	0.36	0.03	0.06
Gas oil and diesel oil (without biofuels)	0.14	0.12	0.01	-	0.11	-	0.02	-	0.00	0.01	0.00	0.00
Fuel oil	0.51	0.46	0.05	0.00	0.40	0.00	0.06	0.00	0.01	0.03	0.00	0.01
Other oil products	2.24	0.03	-	-	0.03	-	2.21	-	-	2.21	-	-
Natural gas	42.30	16.15	0.08	0.16	0.11	15.79	26.16	2.73	1.37	21.61	0.21	0.25
Renewables and biofuels	31.32											
Hydro power	10.91											
Wind power	5.35											
Solar photovoltaic	7.35											
Geothermal	1.87											
Primary solid biofuels	1.27	0.66	0.09	-	0.58	-	0.61	0.17	0.00	-	0.06	0.37
Biogas	2.50	0.89	0.88	0.01	0.00	0.00	1.61	1.60	0.00	0.01	-	0.00
Renewable municipal waste	0.72	0.35	-	-	0.35	-	0.37	0.11	0.00	-	0.04	0.23
Other liquid biofuels	1.34	0.93	0.31	0.00	-	0.61	0.42	0.12	0.00	-	0.04	0.25
Non-renewable waste	0.75	0.37	-	-	0.37	-	0.38	0.11	0.00	-	0.04	0.23
Industrial wastes	0.03	0.02	-	-	0.02	-	0.01	0.00	-	-	0.00	0.01
Non-renewable municipal waste	0.72	0.35	-	-	0.35	-	0.37	0.11	0.00	-	0.04	0.23
Import net	11.38											
other energy sources	0.20											
Electricity GNC	100.00											

Table A3. Future (2030 BASE) Italian electricity mix.

Source	Total %	Only EL %	CI %	TG %	C %	CC %	CHP %	CIC %	TGC %	CCC %	CPC %	CSC %
Coal	9.56	9.56	-	-	9.56	-	-	-	-	-	-	-
Other bituminous coal	9.56	9.56	-	-	9.56	-	-	-	-	-	-	-
Manufactured gases	0.86	-	-	-	-	-	0.86	-	-	0.86	-	-
Blast furnace gas	0.86	-	-	-	-	-	0.86	-	-	0.86	-	-
Oil and petroleum products	1.20	-	-	-	-	-	1.20	-	0.00	1.20	-	-
Other oil products	1.20	-	-	-	-	-	1.20	-	-	1.20	-	-
Natural gas	40.76	15.67	-	-	-	15.67	25.09	3.91	0.61	20.57	-	-
Renewables and biofuels	38.59											
Hydro power	15.00											
Wind power	7.38											
Solar photovoltaic	9.95											
Geothermal	2.07											
Primary solid biofuels	2.10	1.95	-	-	1.95	-	0.15	0.15	-	-	-	-
Biogas	1.67	0.62	0.61	0.01	0.00	0.00	1.05	1.05	0.00	0.00	-	0.00
Renewable municipal waste	0.43	-	-	-	-	-	0.43	0.15	0.00	-	0.06	0.21
Non-renewable waste	0.43	-	-	-	-	-	0.43	0.15	0.00	-	0.06	0.21
Non-renewable municipal waste	0.43	-	-	-	-	-	0.43	0.15	0.00	-	0.06	0.21
Import net	8.40											
CSP	0.20											
Electricity GNC	100.00											

Table A4. Future (2030 INECP) Italian electricity mix.

Source	Total %	Only EL %	CI %	TG %	C %	CC %	CHP %	CIC %	TGC %	CCC %	CPC %	CSC %
Manufactured gases	0.86	-	-	-	-	-	0.86	-	-	0.86	-	-
Blast furnace gas	0.86	-	-	-	-	-	0.86	-	-	0.86	-	-
Oil and petroleum products	1.20	-	-	-	-	-	1.20	-	0.00	1.20	-	-
Other oil products	1.20	-	-	-	-	-	1.20	-	-	1.20	-	-
Natural gas	34.15	2.31	-	-	-	2.31	31.84	1.71	2.15	27.99	-	-
Renewables and biofuels	53.92											
Hydro power	14.60											
Wind power	11.88											
Solar photovoltaic	21.22											
Geothermal	2.09											
Primary solid biofuels	1.90	1.77	-	-	1.77	-	0.13	0.13	-	-	-	-
Biogas	1.71	0.64	0.62	0.01	0.00	0.00	1.08	1.07	0.00	0.00	-	0.00
Renewable municipal waste	0.51	-	-	-	-	-	0.51	0.15	0.00	-	0.06	0.31
Non-renewable waste	0.51	-	-	-	-	-	0.51	0.15	0.00	-	0.06	0.31
Non-renewable municipal waste	0.51	-	-	-	-	-	0.51	0.15	0.00	-	0.06	0.31
Import net	8.46											
CSP	0.89											
Electricity GNC	100.00											

References

1. European Commission. EU Climate Action. Available online: https://ec.europa.eu/clima/policies/eu-climate-action_en (accessed on 20 January 2020).

2. Ministry of Economic Development, Ministry of the Environment and Protection of Natural Resources and the Sea, Ministry of Infrastructure and Transport. Integrated National Energy and Climate Plan. Available online: https://www.mise.gov.it/index.php/it/energia/energia-e-clima-2030 (accessed on 20 April 2020).

3. ISO-The International Organization for Standardization. *ISO 14040:2006: Environmental Management-Life Cycle Assessment-Principles and Framework*; ISO: Geneva, Swtzerland, 2006.

4. Hammond, G.P.; Jones, C.I.; O'Grady, Á. Environmental Life Cycle Assessment (LCA) of Energy Systems. In *Handbook of Clean Energy Systems*; John Wiley and Sons: New York, NY, USA, 2015; pp. 1–26.

5. Girardi, P.; Brambilla, C.; Mela, G. Life Cycle Air Emissions External Costs Assessment for Comparing Electric and Traditional Passenger Cars. *Integr. Environ. Asses.* **2020**, *16*, 140–150. [CrossRef] [PubMed]

6. San Miguel, G.; Cerrato, M. Life Cycle Sustainability Assessment of the Spanish Electricity: Past, Present and Future Projections. *Energies* **2020**, *13*, 1896. [CrossRef]

7. Thonemann, N.; Maga, D. Life Cycle Assessment of German Energy Scenarios. In *Progress in Life Cycle Assessment*; Springer: Cham, Switzerland, 2019; pp. 165–175.

8. Stamford, L.; Azapagic, A. Life cycle sustainability assessment of UK electricity scenarios to 2070. *Energy Sustain. Dev.* **2014**, *23*, 194–211. [CrossRef]

9. Cellura, M.; Cusenza, M.A.; Guarino, F.; Longo, S.; Mistretta, M. Life Cycle Assessment of Electricity Generation Scenarios in Italy. In *Life Cycle Assessment of Energy Systems and Sustainable Energy Technologies*; Springer: Cham, Switzerland, 2018; pp. 3–15.

10. Hertwich, E.G.; Gibon, T.; Bouman, E.; Arvesen, A.; Suh, S.; Heath, G.A.; Shi, L. Integrated life-cycle assessment of electricity-supply scenarios confirms global environmental benefit of low-carbon technologies. In Proceedings of the National Academy of Sciences Annual Meeting, Washington, DC, USA, 25–28 April 2015; Volume 112, pp. 6277–6282.

11. Gibon, T.; Wood, R.; Arvesen, A.; Bergesen, J.D.; Suh, S.; Hertwich, E.G. A methodology for integrated, multiregional life cycle assessment scenarios under large-scale technological change. *Environ. Sci. Technol.* **2015**, *49*, 11218–11226. [CrossRef] [PubMed]

12. Murphy, D.J.; Raugei, M. The Energy Transition in New York: A Greenhouse Gas, Net Energy, and Life-Cycle Energy Analysis. *Energy Technol.* **2020**, 1901026. [CrossRef]

13. Astudillo, M.F.; Treyer, K.; Bauer, C.; Pineau, P.O.; Amor, M. Life cycle inventories of electricity supply through the lens of data quality: Exploring challenges and opportunities. *Int. J. Life Cycle Assess.* **2017**, *22*, 374–386. [CrossRef]

14. Beylot, A.; Guyonnet, D.; Muller, S.; Vaxelaire, S.; Villeneuve, J. Mineral raw material requirements and associated climate-change impacts of the French energy transition by 2050. *J. Clean. Prod.* **2019**, *208*, 1198–1205. [CrossRef]

15. EC-JRC. ILCD Handbook: Recommendations for Life Cycle Impact Assessment in the European Context. 2011. Available online: http://eplca.jrc.ec.europa.eu/uploads/2014/01/ILCD-Recommendation-of-methods-for-LCIAdef (accessed on 20 January 2020).

16. Wernet, G.; Bauer, C.; Steubing, B.; Reinhard, J.; Moreno-Ruiz, E.; Weidema, B. The ecoinvent database version 3 (part I): Overview and methodology. *Int. J. Life Cycle Assess.* **2016**, *21*, 1218–1230. [CrossRef]

17. TERNA. Dati Statistici Sull'energia Elettrica in Italia. 2017. Available online: https://www.terna.it/it-it/sistemaelettrico/statisticheeprevisioni/datistatistici.aspx (accessed on 1 March 2019).

18. Eurostat. Energy Balances April 2019 Edition. Available online: https://ec.europa.eu/eurostat/web/energy/data/energy-balances (accessed on 1 March 2020).

19. SNAM. Bilancio di Sostenibilità. 2014. Available online: www.snam.it (accessed on 1 October 2015).

20. Ministero dello Sviluppo Economico. Produzione Nazionale di Idrocarburi-Anno. 2015. Available online: http://unmig.sviluppoeconomico.gov.it/unmig/produzione/produzione.asp (accessed on 1 October 2015).

21. ISPRA. Fattori di Emissione per la Produzione e il Consumo di Energia. 2017. Available online: http://www.sinanet.isprambiente.it/it/sia-ispra/serie-storiche-emissioni/fattori-di-emissione-per-la-produzione-ed-il-consumo-di-energia-elettrica-in-italia/view (accessed on 15 June 2019).

22. IPCC. Guidelines for National Greenhouse Gas Inventories–Chapter 2–Stationary Combustion. 2006. Available online: http://www.ipcc-nggip.iges.or.jp/public/2006gl/vol2.html (accessed on 2 June 2018).

23. Funedda, M. *Sarlux Srl Dichiarazione Ambientale*; Sarlux Srl: Cagliari, Italy, 2018.

24. European Commission. *EU Reference Scenario (REF2016)*; Publications Office of the European Union: Luxembourg, 2016.

25. Sala, S.; Crenna, E.; Secchi, M.; Pant, R. *Global Normalisation Factors for the Environmental Footprint and Life Cycle Assessment*; Publications Office of the European Union: Luxembourg, 2017. [CrossRef]

26. Ardente, F.C.; Latunussa, E.; Blengini, G.A. Resource efficient recovery of critical and precious metals from waste silicon PV panel recycling. *Waste Manag.* **2019**, *91*, 156–167. [CrossRef] [PubMed]

27. Louwen, A.; Van Sark, W.G.J.H.M.; Schropp, R.E.I.; Turkenburg, W.C.; Faaij, A.P.C. Life-cycle greenhouse gas emissions and energy payback time of current and prospective silicon heterojunction solar cell designs. *Prog. Photovolt. Res. Appl.* **2015**, *23*, 1406–1428. [CrossRef]

28. Santoyo-Castelazo, E.; Azapagic, A. Sustainability assessment of energy systems: Integrating environmental, economic and social aspects. *J. Clean. Prod.* **2014**, *80*, 119–138. [CrossRef]

© 2020 by the authors. Licensee MDPI, Basel, Switzerland. This article is an open access article distributed under the terms and conditions of the Creative Commons Attribution (CC BY) license (http://creativecommons.org/licenses/by/4.0/).

Article

Life-Cycle Carbon Emissions and Energy Return on Investment for 80% Domestic Renewable Electricity with Battery Storage in California (U.S.A.)

Marco Raugei [1,2,3,*], Alessio Peluso [1], Enrica Leccisi [2] and Vasilis Fthenakis [2]

[1] School of Engineering, Computing and Mathematics, Oxford Brookes University, Wheatley, Oxford OX33 1HX, UK; apeluso@brookes.ac.uk
[2] Center for Life Cycle Assessment, Columbia University, New York, NY 10027, USA; el2828@columbia.edu (E.L.); vmf5@columbia.edu (V.F.)
[3] The Faraday Institution, Didcot OX11 0RA, UK
* Correspondence: marco.raugei@brookes.ac.uk

Received: 29 June 2020; Accepted: 19 July 2020; Published: 1 August 2020

Abstract: This paper presents a detailed life-cycle assessment of the greenhouse gas emissions, cumulative demand for total and non-renewable primary energy, and energy return on investment (EROI) for the domestic electricity grid mix in the U.S. state of California, using hourly historical data for 2018, and future projections of increased solar photovoltaic (PV) installed capacity with lithium-ion battery energy storage, so as to achieve 80% net renewable electricity generation in 2030, while ensuring the hourly matching of the supply and demand profiles at all times. Specifically—in line with California's plans that aim to increase the renewable energy share into the electric grid—in this study, PV installed capacity is assumed to reach 43.7 GW in 2030, resulting of 52% of the 2030 domestic electricity generation. In the modelled 2030 scenario, single-cycle gas turbines and nuclear plants are completely phased out, while combined-cycle gas turbine output is reduced by 30% compared to 2018. Results indicate that 25% of renewable electricity ends up being routed into storage, while 2.8% is curtailed. Results also show that such energy transition strategy would be effective at curbing California's domestic electricity grid mix carbon emissions by 50%, and reducing demand for non-renewable primary energy by 66%, while also achieving a 10% increase in overall EROI (in terms of electricity output per unit of investment).

Keywords: grid mix; California; energy transition; life cycle assessment; net energy analysis; EROI; photovoltaic; energy storage; lithium-ion battery; hourly data

---

1. Introduction

Today, ensuring the energy delivery that societies need for productivity, economic growth and well-being is crucial. Since industrialization, the world is experiencing an increase in human population and energy demand per capita, and this has led to a rapid increase in carbon dioxide (CO_2) concentration in the atmosphere from 280 ppm (parts per million) to over 400 ppm. [1]. Curbing a further increase in carbon emissions is one of the major challenges of this century as discussed in the negotiations culminating in The Paris Agreement in 2015 [2]. This agreement to address climate change was the first that was signed by 195 countries, and its overall aim is to keep the global temperature below 2 °C above the pre-industrialization level, encouraging additional efforts to limit global warming to below 1.5 °C. As a consequence, all parties of the United Nation Framework Convention for Climate Change (UNFCCC) act to reduce greenhouse gas (GHG) emissions to the atmosphere though a range of measures, among which decarbonizing electric grid systems plays a prominent role.

Worldwide, increasing and joint efforts have been put to analyse the possible energy transition pathways towards renewable energy sources, assessing their technical feasibility, environmental impacts, and energy implications, dating back to the mid-1970s [3–9].

Assessing the full environmental impacts and the energy burdens of key electricity generation technologies such as solar photovoltaics, wind and nuclear is crucial because although they are almost "zero carbon" at their use-phase, there are still impacts associated to their manufacturing, which should also be taken into the account. A fundamental framework that addresses the cradle-to-grave impacts of human-dominated systems and services is the Life Cycle Assessment (LCA) methodology, which characterises and quantifies all the life-cycle stages from raw material extraction to processing, distribution, manufacturing, operation and decommissioning.

It is also worth noting that it is important to assess each electricity grid as a whole, including all the electricity generation, transmission and storage technologies, and estimate the associated overall environmental impacts and energy implications, as discussed in some recent studies [10–17].

That is because the respective impact of each electricity grid depends on specific conditions such its composition, location, as well as on the local demand profile, and on the required amount (and type) of energy storage.

The state of California in the U.S.A. has established one of the most ambitious plans to bring about an energy transition from fossil fuel generation technologies towards renewable energies, with an intent to generate 60% of its electricity demand using renewable energies by 2030, and 100% thereof using a mix of "zero carbon" sources–including renewables and nuclear—by 2045 [18]. This plan also aims to reach a 40% reduction in GHG emissions below 1990 levels by 2030, and an 80% reduction by 2045.

Such rapid increase of renewable energy penetration in the California electricity grid is expected to require energy storage systems because of the intrinsic intermittency of renewable generation profiles. Specifically, at high penetration, increased photovoltaic (PV) installation is synergistic with storage technologies, which play a critical role in deep decarbonization scenarios, as discussed in recent publications [19–22].

According to a study prepared by the National Renewable Energy Laboratory [23], even with optimal grid improvements, California would still need an estimated 15 GW of additional storage just to reach 50% solar generation by 2030, which is more than 11 times the amount of storage currently mandated in California, and 66 times the total storage power deployed in the U.S. in 2016. This implies that energy storage will continue to be a main ingredient in the mix of strategies to balance supply and demand, support the California Independent System Operator (CAISO) in maintaining grid stability, avoid voltage and frequency imbalances, and support the state's transition to a renewables-centric energy infrastructure [24,25].

For instance, the technical feasibility of utility PV systems plus battery energy storage as an alternative to gas peakers in California is assessed in Roy et al. [26]. Their findings show that a 50 MWAC PV system with 60 MW/240 MWh battery storage can provide more than 98% capacity factor over the target 7:00–10:00 p.m. period, with lower lifetime cost of operation (LCOO) than a conventional combustion turbine natural gas power plant. LCOO includes installed costs, fixed and variable operation & maintenance (O&M), fuel costs as well as other policy factors such as tax credits/incentives.

As highlighted in the recent literature, there is also a need to assess the environmental impacts associated with the increasing energy storage technologies in combination with renewables in the electricity grids in order to better understand and mitigate them [27].

In a recent publication, Raugei et al. [28] estimate the incremental energy and environmental impacts of adding lithium-ion battery (LIB) storage capacity to photovoltaics. Such analysis shows that the energy payback time (EPBT) and life-cycle global warming potential (GWP) increase by 7–30% (depending on storage duration scenarios), with respect to those of PV without storage, and thus the benefits of PV when displacing conventional thermal electricity (in terms of carbon emissions and energy renewability) appear to be only marginally affected by the addition of energy storage.

However, a generalized grid mix was considered in that study, and curtailment and storage figures were assumed without the support of specific historical data. The actual energy and environmental impacts of energy storage in real-world application scenarios will also depend on the specific storage quantities, types and use strategies [29].

In large renewable energy penetration scenarios, there is also a need for analyses at the whole grid level—taking into the account the specific electricity grid mix composition—with an accurate quantification of storage demand and curtailment, which should be informed by detailed hourly generation profiles. It is crucial to identify each hourly mismatch between the demand profile and actual generation provided, especially during peak hours. The importance of such mismatch was also highlighted in 2013 by CAISO in their published chart [30], famously nicknamed "the duck curve", which has since become part of common terminology for describing the effects of large-scale deployment of solar photovoltaic power into the electric grids. The curve shows the difference in electricity demand and the amount of available solar energy throughout, considering a 24-h period in California during springtime.

In light of all of the above, the decision was made here to collect full hourly electricity generation and demand data for California [31], and then use such data as the basis for modelling the amount of energy storage that will be required to minimize the reliance on natural gas and imports when larger quantities of renewables are deployed into the grid. Specifically, the aim of this study is to quantify the life-cycle environmental and energy burdens associated with the current (2018) composition of the electricity grid in California (in terms of greenhouse gas emissions, total and non-renewable cumulative energy demand, and energy return on investment), and compare them to those for a prospective grid mix in 2030, defined so as to achieve 80% of domestic renewable electricity generation, with a suitable amount of storage informed by the detailed hourly generation and demand model.

2. Materials

2.1. Power Dispatch Data for California

2.1.1. Electricity Generation Data per Technology (Hourly Resolution)

The Congress of the United States of America conceived an Open Access Same-time Information System (OASIS) with the Energy Policy Act of 1992, aimed at improving energy efficiency [32], and the Federal Energy Regulatory Commission (FERC) formalised the OASIS with two orders in 1996 [33,34]. Accordingly, CAISO developed an OASIS to provide market observers with easy access to electricity generation historical data [35]. This article is based on historical data for net electricity generation from the OASIS archive, collected separately for each technology and in hourly resolution for the entire baseline year of 2018.

2.1.2. Electricity Imports Data (Hourly Resolution)

The OASIS archive also provides data for electricity power transferred to the California state's grid from other states to satisfy the in-state demand (electricity imports). Imports accounted for 27% of 2018 demand. Such imports are taken into account here for balancing the supply and demand profiles but are excluded from the scope of the environmental and energy assessment, since the latter are focused on the domestic grid mix of the California state.

2.1.3. Electricity Demand Data (Hourly Resolution)

The OASIS-sourced hourly electricity demand data were checked against the corresponding values calculated as the sum of the in-state electricity generation by fuel type plus the electricity imports (all data in matching hourly resolution), to ensure consistency across all datasets used.

2.1.4. Power Curtailment Data for Wind and Solar Generation (Hourly Resolution)

CAISO curtails power flows across the grid system during system emergencies that can affect reliability and safety of the power grid. The operator provides daily reports for the wind and solar electricity generation curtailed specifying the reasons for the curtailment. Reasons of curtailment can be technical or economic, and at either local or system-wide levels, to mitigate congestion, or to mitigate oversupply (defined as when wind or PV facilities deliver more power than is required). An overview of the power curtailment data from the OASIS historical data archive showed that, in 2018, in no case did curtailment occur to mitigate oversupply.

2.2. Current and Future Electricity Generation and Storage Technologies in California

2.2.1. Nuclear

There is only one nuclear power plant currently operating in California, managed by Pacific Gas and Electric Company (PG&E). The plant totals 2393 MW of installed capacity [36] and uses two pressurized-water reactors (PWRs) to generate electricity [37]. This power plant was modelled here using the Ecoinvent process for PWRs in the Western Electricity Coordinating Council (WECC) region, which includes California [38]. It is expected that both reactors will be decommissioned within the time frame of interest for this study, and specifically in November 2024 and August 2025, respectively [39].

2.2.2. Gas-Fired Electricity

In California, natural gas utilities are regulated by the California Public Utilities Commission (CPUC) and those utilities are managed by several service providers such as Pacific Gas and Electric Company (PG&E), Southern California Gas Company (SoCalGas), San Diego Gas & Electric Company (SDG&E), Southwest Gas. At present, Natural Gas Combined-Cycle (NGCC) plants represent the 41% of the total 42,695 MW installed capacity, and 69% of the total electricity generated by natural gas-fired plants [40]. According to the 2018 California Gas Report [41], gas demand for electricity generation is expected to decline due to California's programs to minimise greenhouse gas (GHG) emissions, with a concomitant increase in renewable energy (RE) technologies. In this study, the life-cycle inventories for NGCC and single-cycle gas turbines (SCGT) operated in California were based on the corresponding Ecoinvent processes for the WECC region [38].

2.2.3. Geothermal

California is located within the "Pacific ring of fire" geographic area, where the frequency of earthquakes and volcanic eruptions are highest as a result of the movement of tectonic plates. California is also characterised by a large number of natural geysers which provide a natural resource of geothermal power. There are 43 geothermal power plants in California, totalling 2730 MW of installed capacity [36], which use natural steam to drive turbines which in turn are used as generators to produce electricity for the grid. The life-cycle inventory for geothermal electricity was based on the corresponding Ecoinvent processes for the WECC region [38].

2.2.4. Biomass

Biomass power plants in California use a combination of raw biomass residues (including forest and agricultural residues from trees, foliage, roots and chips from wood processing residues) and, secondarily, municipal solid waste residues (mostly cellulose) [42]. A total installed capacity of 1325 MW is reported [36], which includes biogas (*cf.* Section 2.2.5). The WECC heat and power co-generation model from the Ecoinvent database [38] was selected to represent biomass electricity generation, and all energy and environmental impacts of this multi-output process were allocated on an energy content basis. This model is limited to the use of woodchips as feedstock; however, given that biomass-fired plants contribute just 2% to the total electricity generated in-state (see Section 2.3), such simplification was deemed acceptable.

2.2.5. Biogas

Biogas is a mixture of gases produced from the anaerobic decomposition of municipal solid waste (kitchen waste, garden waste), livestock manure, food processing waste, agricultural wastes and industrial wastewater. The process occurs in a controlled environment such as in airtight containers, in floating covers on lagoons or directly in landfills [43].

The resulting biogas is mostly composed of methane (CH_4) and carbon dioxide (CO_2), but also contains other hydrocarbons and significant traces of ammonia (NH_3) and hydrogen sulfide (H_2S).

The Ecoinvent WECC biogas-fired heat and power co-generation model [38] was selected to assess biogas electricity generation, and all energy and environmental impacts of this multi-output process were allocated on an energy content basis.

2.2.6. Hydro

Hydroelectric power in California can be divided into two different categories. There are facilities which use a dam to create a reservoir to generate electricity, with an installed capacity that is typically larger than 30 MW, and there are other facilities which divert water from a river or stream to generate electricity, with typically much smaller unit power capacities. In 2018, the former totalled 12,281 MW and generated 13% of the total in-state electricity output, while the latter clocked in at just 1758 MW and 2% of generated electricity [36]. The corresponding Ecoinvent processes for dammed reservoir and run-of-river hydroelectricity in the WECC region [38] were adopted here.

2.2.7. Wind

Wind power generation in California has a long history dating back to 1980, and current total installed capacity is 5964 MW [36]. All wind farms in California are on-shore, with different types of generators, ranging from the older ones with a typical installed capacity of less than 1 MW, to some recent ones with capacities of over 3 MW. Currently, the U.S. Wind Turbine Database of the U.S. Geological Survey [44] estimates the number of generators installed in California to be approximately 6000 units, of which slightly less than 3000 units are characterized by an installed capacity between 1 MW and 3 MW. However, the database includes decommissioned and duplicate turbines; combined with the uncertainty on specific technical data for the turbines, it was thus deemed acceptable to adopt the Ecoinvent process for 1–3 MW onshore wind turbines in the WECC region as the most representative proxy for the totality of the wind farms in California. The model assumes a 20-year lifetime for all moving components and a 40-year lifetime for all the stationary components of the wind installation [38].

2.2.8. Concentrating Solar Power (CSP)

Concentrating solar power (CSP) generation in California has a long history dating back to 1984. All CSP technologies entail a transfer fluid which absorbs the sun's energy and is used to heat water and produce steam, which then drives a steam turbine generator. There are four CSP technologies, which differ in terms of the receiver system: solar towers, parabolic troughs, compact linear Fresnel reflectors, and dish engines. In California, the total installed CSP is 1249 MW [36], and 69% of the CSP electricity is generated by parabolic troughs, and the remaining 31% by solar towers [45]. Parabolic trough systems focus solar radiation onto a receiver tube that runs down the center of a trough by using curved mirrors; while solar tower systems focus solar radiation on a receiver at the top of a high tower by using computer-controlled mirrors, called heliostats, which track the sun along two axes. The corresponding Ecoinvent processes for parabolic trough and solar tower installations [38] were checked, but the data are specific to South Africa, and they do not consider the production of solar salts (the latter are expected to be added in the next release of the database) [46]. Therefore, data from the open literature were used for CSP because of their representativeness and completeness, specifically: a wet-cooled 103 MW parabolic trough concentrating solar power (CSP) located in Daggett, CA [47] and a dry-cooled, 106 MW power tower CSP located near Tucson, AZ. Both systems use a mixture of mined nitrate salts for energy storage [48].

2.2.9. Photovoltaic (PV) Solar

In California, solar PV systems have been growing rapidly over the years due to a favourable combination of high insolation, community support, and declining PV panel costs. Currently, the California electricity grid features both utility-scale and distributed rooftop solar PV systems, totalling 10,661 MW [36]. Considering the specific topography, the land availability, and California's plan to increase the PV penetration in its electric grid, it is reasonable to assume that utility-scale PV installations in particular will continue to expand the most in future years.

For the purposes of this analysis, then, utility-scale PV installations were assumed throughout, with panel shares corresponding to 33% single-crystalline silicon (sc-Si), 62% multi-crystalline silicon (mc-Si), and 5% cadmium telluride (CdTe), which reflect the current global production data collected in the latest Fraunhofer Institute for Solar Energy report [49]. The assumed energy capture efficiency of each PV panel type is also based on the same Fraunhofer report which provides the current commercial average efficiencies in 2018 [49], namely: 18% for sc-Si, 17% for mc-Si, and 18% for CdTe.

In order to model the PV systems, the latest available foreground inventory data were used, as discussed in a previous paper [50]. Specifically, for c-Si PV modules, the foreground inventory data source was the latest IEA-photovoltaic power systems (PVPS) Task 12 Report [51]. For CdTe PV modules, up-to-date production data were provided directly by First Solar, which is currently the leader producer for this technology. The same company also provided information on the balance of system (BOS) for typical ground-mounted installations, which was also adopted for the c-Si technologies.

The main background data source was the Ecoinvent database [38], but all the in-built assumptions were adapted to the current production conditions in order to be as accurate and realistic as possible. Specifically, the main producer country for c-Si PV panels is now China, while CdTe is mainly produced in the US and in Malaysia. Accordingly, the corresponding electricity generation mixes were used to model the production of the PV modules.

End-of-life (EoL) management and decommissioning of the PV systems were not included in this analysis for consistency with other grid technologies in the analysis. However, it is worth noting that including EoL may actually provide environmental and economic benefits, due to the possible recycling of the components, especially aluminium and silicon [52]. Also, metal recycling—such as the copper contained not only in the PV panels, but also in the BOS—could be strategic in order to further reduce the environmental impacts of PVs.

Finally, given that PV systems are still on a continuously and rapidly improving trend, their expected future efficiencies were estimated on the basis of recent IEA projections [53]. Specifically, for 2030 the following conservative efficiencies improvements were assumed: 21% for sc-Si and CdTe, and 20% for mc-Si. All lifetimes were kept constant at the industry-standard of 30 years. A second, more aggressive efficiency improvement trajectory was also considered by way of sensitivity analysis, whereby 23% efficiencies were set for sc-Si and CdTe, and 22% for mc-Si, coupled with improved 40-year lifetimes [53]. Both future projections for PV may still be considered conservative, however, since all other modelling parameters (including photoactive layer thickness, material usage efficiency and foreground energy inputs per m^2 of PV module) were kept constant in all cases. Additionally, next-generation PV technologies (e.g., single-junction and tandem perovskites) may become viable in the medium-term future which could reduce the energy and environmental impacts of PV electricity even further [54].

2.2.10. Energy Storage

According to the literature, there are six main types of technologies which can provide energy storage, namely electrochemical, mechanical, gravitational, chemical, thermal and electrical storage [55].

Currently, in most cases the balance and the flexibility for a power grid is entrusted to pumped hydro storage (PHS) as the primary choice, when possible, due to its long technical lifetime and generally low economic, energy and environmental impacts. However, it is expected that electrochemical storage will play an increasingly important role in the next future, when more storage capacity will be required because of increased penetration of variable renewable energy (VRE). Specifically, lithium-ion batteries (LIB) are considered the most likely candidates for reasons of expected cost reductions [56], charge capability, energy density and efficiency [55,57].

For the purposes of this analysis, it was assumed that 100% of the required storage capacity to balance the California grid in the analysed 2030 scenario will be provided by stationary installations of LIBs. The main reason for this assumption is that this article aims to provide a conservative (i.e., worst case) analysis which excludes any opportunity to resort to using existing in-state or out-of-state PHS to provide part of the storage requirement. Additionally, it is acknowledged that in actuality other forms of storage, such as small-scale off-river pumped hydro [58] and compressed air, could be deployed alongside LIBs, thereby further reducing the demand for natural gas, potentially even to zero. However, such additional storage options and even more aggressive energy storage deployment fall outside of the scope of this study.

LIB storage was modelled on the basis of the Ecoinvent model for lithium manganese oxide (LMO) technology [38]. Round-trip storage efficiency was set at 80% [59], and the expected service lifetime of the batteries was conservatively set at 7000 cycles (corresponding to a residual depth of discharge of 80% for LMO technology) [60]. A previous study on PV + LIB storage [28] performed a sensitivity analysis whereby LMO batteries were compared to nickel-cobalt-manganese (NCM) and lithium-iron phosphate (LFP) alternatives, but the results showed comparatively small variance ranges for both energy and greenhouse gas impacts. Conservatively, no improvements in energy storage density, material usage efficiency or foreground energy inputs to LIB production were considered for 2030, relative to the present.

2.3. California Grid Mix Composition in 2018

In 2018 the total California domestic generation was 165 TWh. Figure 1 illustrates the California domestic grid mix composition in terms of total in-state electricity generated in the year 2018. Eleven % of the total in-state electricity was supplied by nuclear reactors, but as explained in Section 2.2.1, all of this is expected to be completely phased out by 2025.

Gas-fired electricity represented 39% of the total in-state generation, but that too is expected to decline due to aggressive California programs to minimize greenhouse gas (GHG) emissions. At the same time, though, it is also expected that gas-fired generation will continue to be a valuable technology for load following, and to compensate for the intermittency of wind and solar generation.

The remaining 50% of the total in-state electricity generation was supplied by RE technologies. Specifically, wind installations generated 10% of the total in-state electricity, while PV systems generated 16% thereof. As discussed in Section 3.1, the share of RE, and specifically PV, is expected to increase significantly over the coming years, with a concomitant surge in the required energy storage capacity.

Finally, electricity transmission was also included within the boundary of this assessment, albeit limited to the high voltage (HV) network. This was deemed an acceptable simplification, since the vast majority of the electricity generation plants comprising the grid mix at present and in the considered future scenario are centralised units which inject HV electricity into the grid. The HV transmission lines were modelled using the WECC-specific life-cycle inventory (LCI) information provided in the Ecoinvent database [38], and transmission losses were set at 6% as per historical data [61].

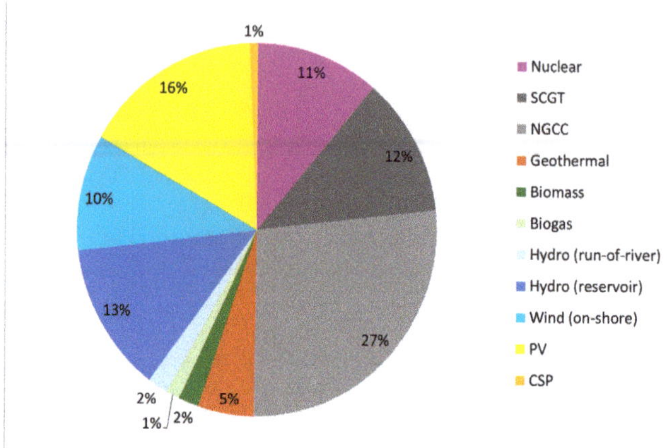

Figure 1. California domestic electricity generation mix—historical data for 2018 from CAISO. Total domestic generation was 165 TWh. SCGT = single cycle gas turbines; NGCC = natural gas combined cycles; PV = photovoltaics; CSP = concentrating solar power.

3. Methods

3.1. Definition of the Future Grid Mix Scenario in 2030

The modelling of the future scenario for the California grid mix in the year 2030 was carried out as described below. Firstly, full hourly-resolution net generation profiles were generated for each technology using the OASIS data collected for 2018 (i.e., the latest available complete datasets at the time of writing). Figure 2 shows the resulting stacked contributions to the total delivered power (black line, which is equal to the demand profile) for a typical day in spring (2nd April). From bottom to top: (I—purple) nuclear (which as expected is almost constant, i.e., a "baseload" provider"), (II—green) other renewables (i.e., the sum of hydro, biogas, biomass and geothermal), (III—blue) wind, (IV—yellow) solar (PV + CSP). Lastly, at any hour, the gap (grey arrow) between the top-most reported generation profile and the demand profile is supplied by a combination of natural gas generation (SCGT + NGCC) and electricity imports.

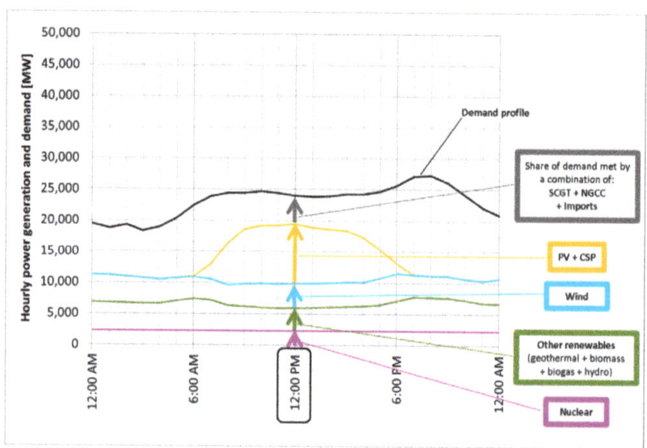

Figure 2. Historical hourly generation and demand profiles for the day of 2 April 2018 in California, from CAISO data.

In addition, the 2018 hourly "potential" PV output was also calculated, by adding back the reported 2018 hourly PV curtailment data to the corresponding net PV generation data. The purpose of this "potential" PV output profile was to provide the basis for the future extrapolation of the corresponding "potential" PV output profile for 2030 (see point 2 below), under the assumption that by then, the large-scale availability of energy storage would negate the need for all technical or economic curtailment other than that due to oversupply (see Section 2.1.3).

Starting from these historical power generation profiles, the corresponding projected profiles for the year 2030 were calculated, based on the modelling assumptions and calculations described below:

(1)   It is assumed that in 2030 the total hourly electricity demand profile will remain the same as in 2018. This future extrapolation is based on the analysis of the demand profiles from 2001 to 2019, which shows that the cumulative yearly electricity demand remained nearly constant during the past 19 years, with minimal oscillations around a centre value of approximately 200 TWh/year. This result appears to be "due to a combination of energy efficiency measures and less electricity-intensive industry that counterbalances increased population and economy" [62]. Other potential variations in electricity demand (both its hourly profile and total year-end cumulative value), for instance due to a possible large-scale deployment of electric vehicles (EVs) and the associated requirement for battery charging, are outside the scope of this study.

(2)   CAISO will rely single-handedly on solar PV as the technology of choice to increase the penetration of renewable energy in the grid. This is a bold assumption, but it was deemed reasonable in view of the abundance of solar irradiation in California, and it also appears to be supported by a simple linear extrapolation of recent past trends, which indicate that wind installations in California have plateaued, whereas PV installations have been sharply and consistently rising (see Figure 3). The final value of installed PV power in 2030 was determined iteratively, so as to match a target of 80% total net domestic renewable electricity generation, after duly taking into account all PV storage and curtailment losses (as explained below). Hence, the PV installed capacity in 2030 is 43,710 MW, as shown in Figure 3. The hourly "potential" (i.e., pre-curtailment and pre-storage) PV output profile was calculated by scaling up the corresponding 2018 "potential" PV electricity generation, proportionally to the respective 2030 vs. 2018 installed power levels.

(3)   Lithium-ion batteries (LIBs) will be deployed as the storage technology of choice (as discussed in Section 2.2.10). The amount of assumed installed LIB power (P) and the maximum consecutive hours of storage duration at such maximum power (t) were set after performing a parametric investigation of the resulting % of VRE curtailment ensuing from a range of P and t values (following the grid balancing algorithm described at points 7 and 8 below). The results of this parametric analysis are illustrated in Figure 4; in order to make a realistically conservative assumption on the amount of storage, for the purposes of this study the choice was therefore made to set P = 60% of the installed PV power (a value consistent with previous literature [63,64]) and t = 6 h. When taken together, such values of P and t lead to the total installed storage capacity E = P × t. As reported in Table 1, this resulted in 2.8% of the overall "potential" VRE generation being curtailed.

(4)   Nuclear generation will be zero, consistently with the planned decommissioning of all remaining reactors in California (as explained in Section 2.2.1).

(5)   "Other renewables" (i.e., hydro, biogas, biomass and geothermal), wind and CSP generation profiles will remain exactly the same as in 2018.

(6)   Single-cycle gas turbines (SCGT) will be completely phased out.

(7)   Combined cycle gas turbine (NGCC) output and electricity imports will be used, together with LIB energy storage, to balance overall supply and demand, following a strict order of merit, as follows:

(a)   On an hourly basis, the increased PV output in 2030 with respect to 2018 (more precisely: the difference between the "potential" PV output in 2030, calculated as per point 2 above, and the net PV output in 2018) will first be compensated for by reducing NGCC output. This is deemed the preferred strategy since gas-fired electricity is the most carbon-intensive technology in the California grid mix, and it is also more carbon-intensive than the average mix of technologies used to generate the electricity imported by California [65].

(b) Then, if/when no residual NGCC power is left, the second intervention will be to curb imported electricity.

(c) Then, if/when the hourly imported electricity value has been reduced to zero too, and the "potential" PV output is actually in excess of the total demand profile value, such excess PV output will be preferentially routed into storage, as long as neither total storage capacity (E) nor maximum storage power (P) are exceeded.

(d) Finally, if, after taking steps (a–c) above, either the maximum E or maximum P condition is met, then the residual excess PV output (i.e., the share thereof that cannot be sent to storage) is curtailed.

(8) After each PV "peak", i.e., as soon as the "potential" PV profile curve has returned below the total demand profile curve, the electricity stored in LIBs will start being dispatched back to the grid (at a maximum rate limited by P), and will thus curb NGCC output (in the first instance) and imported electricity (if/after NGCC output has already been reduced to zero) with respect to their respective 2018 hourly values.

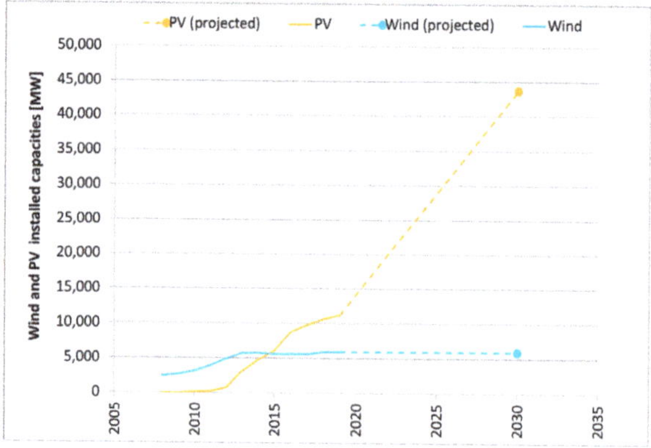

Figure 3. Wind and PV installed capacities in California—historical data from CAISO to 2019 and authors' projections to 2030.

Table 1. Key California ISO (CAISO) grid mix parameters for years 2018 (historical data) and 2030 (projected).

	2018 Grid [%]	2018 Grid [TWh/yr]	2030 Grid [%]	2030 Grid [TWh/yr]
Share of total California demand supplied by domestic generators [1]	73%	165	88%	199
Share of net renewable energy (RE [2]) in domestic generation mix	50%	82	80%	159
Share of net variable renewable energy (VRE [3]) in domestic generation mix	27%	44	61%	121
Share of net PV generation in domestic generation mix	16%	27	52%	104
Share of gross VRE generation that is routed into storage	0%	0	25%	32
Share of gross VRE generation that is curtailed	1%	0.4	2.8%	3.6

[1] Assuming that the total yearly gross electricity demand (pre-transmission losses) remains the same, i.e., 226 TWh/yr.
[2] RE includes: Geothermal, Biomass, Biogas, Hydro, Wind, PV, and CSP. [3] VRE includes: Wind, PV, and CSP.

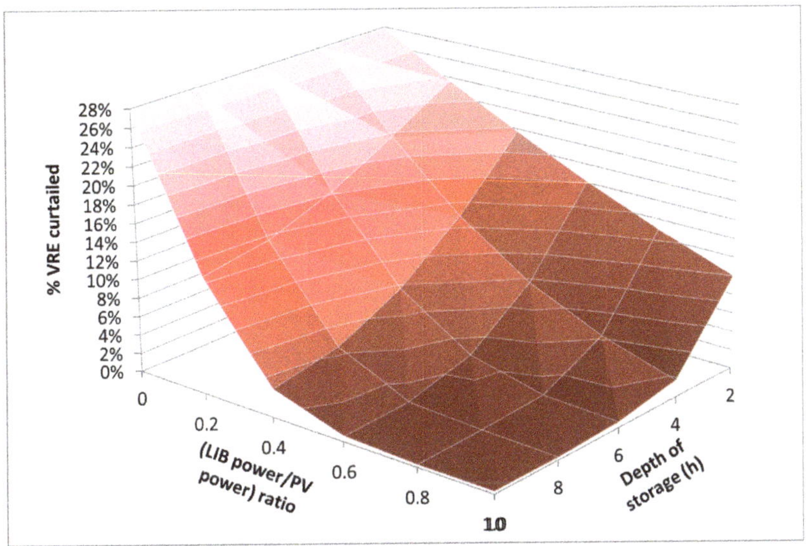

Figure 4. Parametric investigation of variable renewable energy (VRE) curtailment accounting for combinations of lithium-ion battery (LIB) power capacities per PV power and depth of storage (h) in California in the year 2030.

Figure 5 shows the expected demand profile (black line) for 2 April 2030, and the following stacked power generation profiles, from bottom to top: (I—green) other renewables (hydro + biogas + biomass + geothermal), (II—blue) wind, and (III—yellow) solar ("potential" PV + CSP).

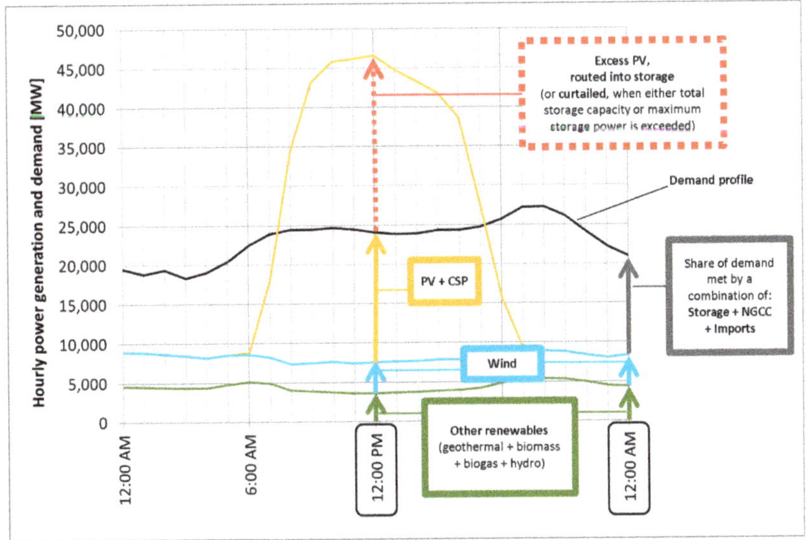

Figure 5. Projected hourly generation and demand profiles for the day of 2 April 2030 in California.

The complete projected hourly electricity generation and demand profiles for the entire year 2030, broken down by month, are reported in the Supplementary Material (Figures S1–S12). Interestingly, because of the large demand for air conditioning in the hotter months in California, the most severe mismatch between the "potential" solar electricity generation and electricity demand profiles occurs in spring, and not in summer, when solar irradiation is highest.

Using the dynamic modelling approach described above, an overall projected year-end domestic grid mix can then be calculated for the California state in 2030. This is illustrated as a pie chart in Figure 6, and its most salient characteristics are compared to those of the corresponding 2018 domestic grid mix in Table 1.

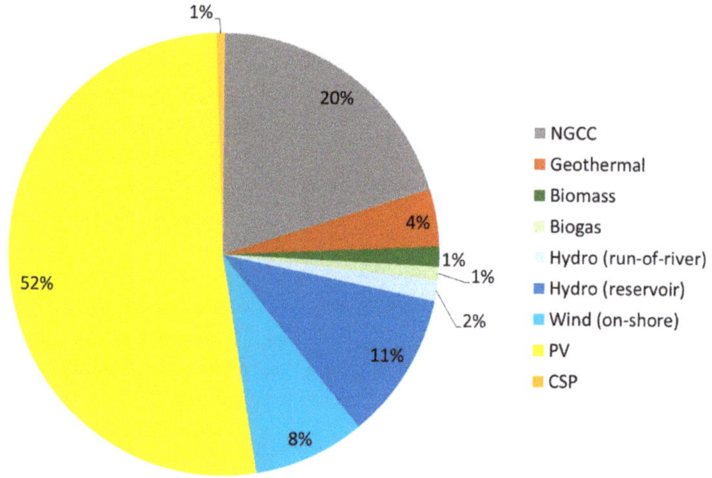

Figure 6. California domestic electricity generation mix—projected data for 2030. Total domestic generation is expected to be 199 TWh. NGCC = natural gas combined cycles; PV = photovoltaics; CSP = concentrating solar power.

The modelled 2030 California grid mix is characterized by a large share of VRE, out of which 25% is not consumed directly but is instead routed into storage, while only 2.8% is curtailed; the resulting share of net (i.e., post-curtailment and storage) VRE in the domestic grid mix is thus 61%. It is also noteworthy that, even after completely phasing out SCGTs, the remaining required NGCC output is also reduced by 30% (relative to 2018). Also, the hypothesised large deployment of PV + storage yields a surplus of available renewable energy, which, when retrieved from storage, allows a significant reduction in electricity imports, and a corresponding surge in the domestic share of total electricity supply in California, from 73% in 2018 to 88% in 2030.

Lastly, a further interesting finding ensued from a separate sensitivity analysis on the key model assumptions and parameters. While, as mentioned in Section 2.1.1, all nuclear capacity is expected to be phased out in California by the mid-2020s, with no plans for new replacement reactors, it was deemed worthwhile to investigate the theoretical effect that retaining the existing nuclear capacity would have on grid stability, demand for storage and corresponding % VRE curtailment. A first alternative grid model run was then carried out for 2030, with the exact same PV and storage capacities as described above, but in the presence of the same nuclear electricity output as in 2018. This resulted in an increased VRE curtailment rate of 4.3%. Such result was found to be due to the inflexibility of nuclear output, which pushed the "potential" PV output peaks even higher with respect to the demand profile, thereby saturating the available storage capacity sooner. In an alternative model run, the storage duration was then adjusted upwards, so as to increase the total storage capacity and thus bring the % VRE curtailment back down to the same 2.8% as in the "baseline" scenario. This ended up requiring the deployment of 7.3 h of LIB storage vs. 6 h in the "baseline" scenario without nuclear electricity in the mix. The details of this analysis are shown in the Supplementary Material (Table S1 and Figure S13).

3.2. Life Cycle Assessment (LCA)

Life cycle assessment (LCA) is a de-facto standard method for the evaluation of the environmental performance of a wide range of industrial processes and technologies, and it mainly owes its wide acceptance to its comprehensiveness (in terms of considering all supply chain stages, from extraction of raw materials through transportation and manufacturing, to use phase and end-of-life). It also benefits from a high degree of standardization [66,67], and from the availability of extensive, industry-vetted inventory databases, among which a prominent role is played by Ecoinvent [38].

Various life-cycle impact assessment methods have been developed, which enable the calculation of dedicated impact indicators for a wide range of impact categories. Among the latter, the focus of this paper is on global warming potential (GWP), estimated using IPCC-derived characterization factors with a time horizon of 100 years (in units of kg of CO_2-equivalent) for all gaseous emissions, excluding biogenic CO_2.

Additionally, two life-cycle energy metrics are also calculated here, namely the cumulative energy demand (CED) and the non-renewable cumulative energy demand (nr-CED), respectively quantifying the total amount of primary energy directly and indirectly harvested from the environment per unit of electricity output, and the non-renewable share thereof (in both cases the results are expressed in MJ of oil-equivalent) [68].

Based on the definition above, it is also self-evident that the life-cycle primary-to-electric energy conversion efficiency of the grid mix taken as a whole (η_G) can be conveniently calculated as the reciprocal of its CED (Equation (1)):

$$\eta_G = 1/CED_G \tag{1}$$

3.3. Net Energy Analysis (NEA)

Net energy analysis (NEA) [69] provides an alternative viewpoint on the energy metabolism of energy harvesting and conversion technologies, whereby the primary energy resource(s) that are directly exploited and converted to useful energy carriers (e.g., the natural gas that is extracted, conveyed by pipeline and then burnt in a power plant to produce electricity; or the solar energy that is harvested and converted to electricity by PV panels) are deliberately excluded from the accounting, and instead the focus is put solely on how much previously-available commercial energy needs to be "invested" in order to operate the energy supply chains (e.g., the energy needed to extract the gas from the ground, build the pipeline, pump the gas through the pipeline, and build the gas turbine; or the energy needed to manufacture the PV panels and their balance-of-system).

When put in rather blunt but arguably vivid terms, it can therefore be said that instead of being concerned with the overall thermodynamic efficiency of a process, NEA aims to quantify the energy "bang for the buck" from the point of view of the end user. Fittingly, its main indicator is the energy return on (energy) investment [70] (defined as per Equation (2)):

$$EROI = Out/Inv \tag{2}$$

However, the NEA literature has historically been characterized by a much lower degree of standardization than the LCA one, which has led to many inconsistent comparisons [71,72].

In this study, in order to integrate the LCA and NEA viewpoints, and to maximize the consistency of the calculations, both internally and externally with some of the more recent literature [10,12,13,15,17], when calculating the EROI of electricity (either produced by a specific technology, or by the grid mix as a whole), all energy investments at the denominator are always accounted for in terms of their respective life-cycle CED (and are thus quantified in units of oil-equivalent).

Then, when the EROI numerator is simply measured as the amount of electricity delivered (i.e., not converted to some form of "equivalent" primary or thermal energy), a subscript "el" is appended to the resulting indicator (i.e., $EROI_{el}$). Alternatively, when the EROI numerator is expressed as "primary energy equivalent" (on the basis of the life-cycle primary-to-electric energy conversion efficiency of the grid mix in the current year), a subscript "PE-eq" is used (i.e., $EROI_{PE\text{-}eq}$).

4. Results and Discussion

Figure 7 illustrates the calculated GWP of the domestic grid mix in California, respectively in 2018 (based on historical data) and in 2030 (when the future grid mix is modelled as described in Section 3.1). The first and foremost result is that the carbon intensity of electricity is expected to be almost halved over the course of a single decade. Such remarkable drop is almost entirely due to the combination of two key factors. Firstly, the massive deployment of PV and energy storage allows a substantial phasing out of gas-fired electricity (and SCGTs in particular). Secondly, the up-front carbon emissions due to the manufacturing and installation of the PV and LIB systems are low enough that, when discounted over the total amount of electricity that they deliver in their combined service lives, they result in comparatively negligible GWP contributions to the grid mix.

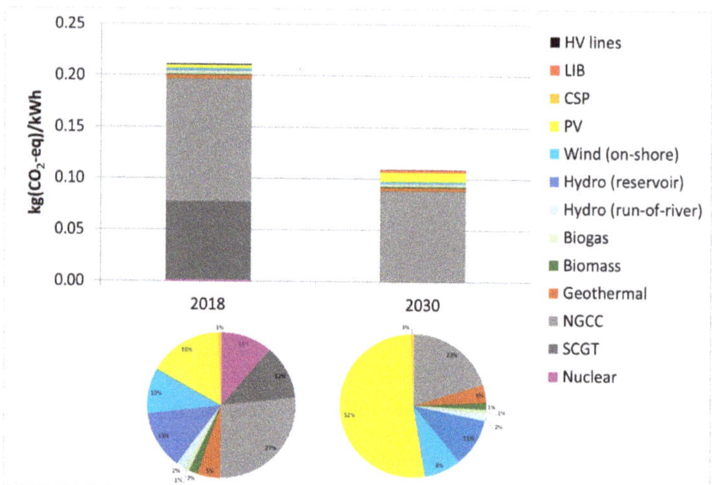

Figure 7. Global warming potential (GWP) results for California domestic grid mix in 2018 and in 2030. The pie charts underneath each bar refer to the corresponding grid mix composition, and are included to aid the interpretation of the results. SCGT = single cycle gas turbines; NGCC = natural gas combined cycles; PV = photovoltaics; CSP = concentrating solar power; LIB = lithium-ion batteries; HV = high voltage.

These results are put in even starker relief when considering that in 2018, gas-fired power plants generated 29% of total domestic electricity while being responsible for 93% of the grid's GWP; conversely, in 2030 PV + LIBs are expected to generate 52% of total domestic electricity while only causing 10% of the grid's total carbon emissions.

In terms of life-cycle energy results, the same planned energy transition results in an overall 31% reduction in the CED of domestic electricity in California (Figure 8), and a corresponding increase in the life-cycle primary-to-electric energy conversion efficiency (η_G) of the grid mix, from 48% to 69%.

The improvement becomes even more significant when specifically focusing on the life-cycle demand for non-renewable primary energy (Figure 9), given that most of the primary energy harvested from the environment to power the grid mix in 2030 is actually renewable (i.e., solar, and to a lesser extent wind, hydro, geothermal and biomass).

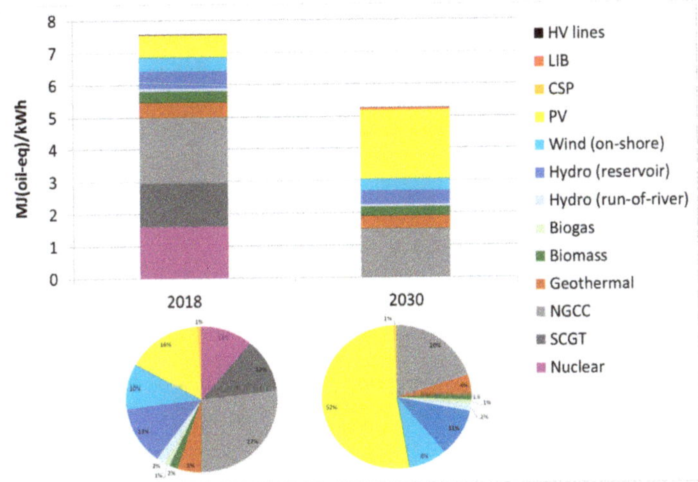

Figure 8. Cumulative energy demand (CED) results for California domestic grid mix in 2018 and in 2030. The pie charts underneath each bar refer to the corresponding grid mix composition, and are included to aid the interpretation of the results. SCGT = single cycle gas turbines; NGCC = natural gas combined cycles; PV = photovoltaics; CSP = concentrating solar power; LIB = lithium-ion batteries; HV = high voltage.

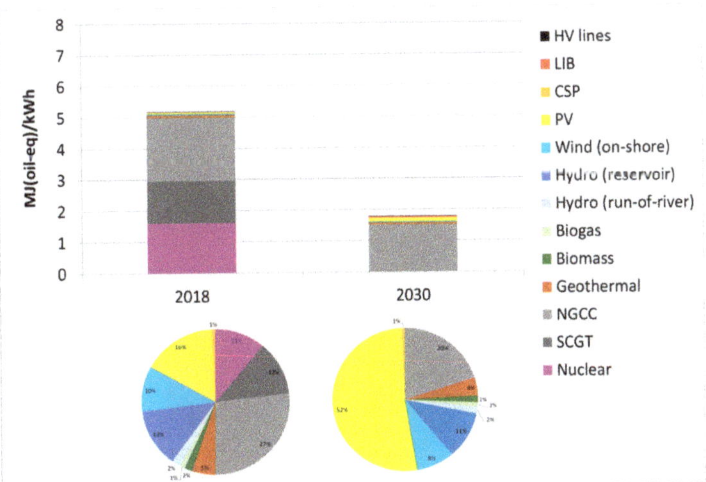

Figure 9. Non-renewable cumulative energy demand (nr-CED) results for California domestic grid mix in 2018 and in 2030. The pie charts underneath each bar refer to the corresponding grid mix composition, and are included to aid the interpretation of the results. SCGT = single cycle gas turbines; NGCC = natural gas combined cycles; PV = photovoltaics; CSP = concentrating solar power; LIB = lithium-ion batteries; HV = high voltage.

As a result, the nr-CED of domestic electricity in California drops by a factor of three, from 5.2 to 1.8 MJ(oil-eq)/kWh. To this effect, it is noteworthy that phasing out nuclear, as well as natural gas, is also beneficial (while nuclear energy is a low-carbon technology, it still obviously relies on non-renewable stocks of fissile fuel, which are also not available domestically in California, adding further meaning to these results in terms of improved energy sovereignty).

Finally, when shifting the viewpoint to the one characteristic of NEA, and thus focusing only on the energy investment per unit of electricity delivered, while excluding the primary energy that is directly harvested and converted to electricity, Figure 10 shows that the planned massive deployment of PV and LIB storage in 2030 does result in significant shares of the total grid mix energy investment being required for these technologies (respectively, 36% and 9%). Even so, the overall energy investment per unit of delivered electricity in 2030 is still reduced with respect to the historical value for 2018.

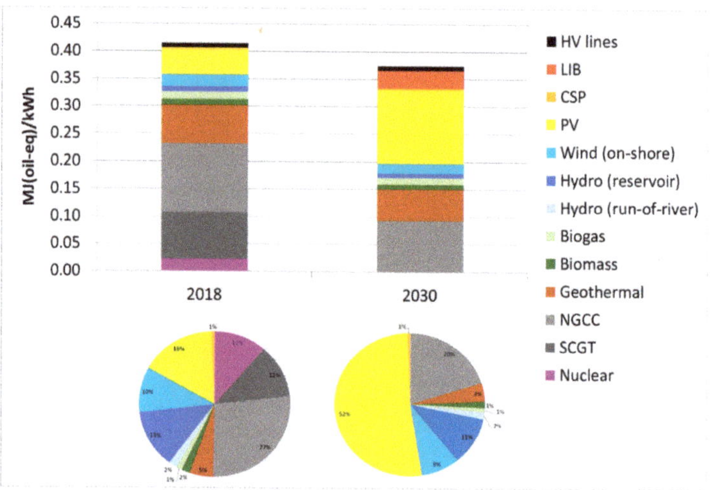

Figure 10. Primary energy investment results for California domestic grid mix in 2018 and in 2030. The pie charts underneath each bar refer to the corresponding grid mix composition, and are included to aid the interpretation of the results. SCGT = single cycle gas turbines; NGCC = natural gas combined cycles; PV = photovoltaics; CSP = concentrating solar power; LIB = lithium-ion batteries; HV = high voltage.

As illustrated in Figure 11, this results in a 10% increase in the $\mathrm{EROI_{el}}$ of the California domestic grid mix as a whole. At the same time, however, because of the larger penetration of PV and the phasing out of nuclear and, partially, gas-fired electricity, the life-cycle primary-to-electric energy conversion efficiency of the grid mix (η_G) increases by as much as 44% in relative terms, from $\eta_G = 0.48$ in 2018 to $\eta_G = 0.69$ in 2030. Therefore, the trend in $\mathrm{EROI_{PE\text{-}eq}} = \mathrm{EROI_{el}}/\eta_G$ ends up being dominated by the latter change in η_G.

In order to provide additional detail on these NEA calculations, the specific EROI results (in terms of both electricity and equivalent primary energy) for the individual electricity generation technologies comprising the California domestic grid mix in 2018 and 2030 are reported in Table 2. Once again, it is noteworthy that the η_G values for California in 2018 and 2030 are significantly higher than typically assumed for electricity grids with higher percentages of thermal technologies ($\eta_G = 0.30$–0.35), resulting in comparatively lower values of $\mathrm{EROI_{PE\text{-}eq}}$. This showcases how any specific $\mathrm{EROI_{PE\text{-}eq}}$ values are only valid for the actual conditions considered in each study (such as grid mix composition, year, and location).

Specifically, changes in η_G are at the root of the differences in $\mathrm{EROI_{PE\text{-}eq}}$ results for PV in California vs. those previously reported by the same authors when considering a more generalised thermal grid mix ($\eta_G = 0.30$) [50].

Finally, by way of sensitivity analysis, an alternative scenario for 2030 was also analysed, in which more efficient and longer-lasting PV systems were assumed (*cf.* Section 2.2.9). The ensuing variations in the calculated energy and carbon emission indicators for the California domestic grid mix are reported in Table 3. As can be seen, this sensitivity analysis proves that the main results of this study are very robust and not likely to be affected significantly by alternative future PV developments.

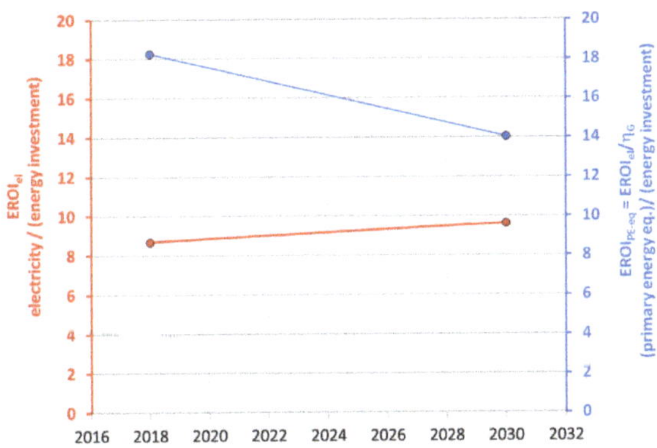

Figure 11. Energy return on (energy) investment ($EROI_{el}$ and $EROI_{PE\text{-}eq}$) of the California domestic grid mix.

Table 2. $EROI_{el}$ and $EROI_{PE\text{-}eq}$ values for individual electricity generation technologies in California, in the years 2018 and 2030.

Technology	2018 $EROI_{el}$	2018 $EROI_{PE\text{-}eq}$ ($\eta_G = 0.48$)	2030 $EROI_{el}$	2030 $EROI_{PE\text{-}eq}$ ($\eta_G = 0.69$)
Nuclear (pressure water reactor)	20	42	N/A	N/A
Natural gas (single-cycle gas turbines)	5	11	N/A	N/A
Natural gas (combined cycles)	8	17	8	12
Geothermal [a]	3	6	3	4
Biomass (co-generation) [a]	6	14	6	9
Biogas (co-generation) [a]	4	7	4	5
Hydro (run-of-river)	70	148	70	102
Hydro (reservoir)	53	112	53	78
Wind (on-shore)	18	37	18	25
Photovoltaic	13	28	15 [b]	22 [b]
Concentrating solar power	8	18	8	12

[a] EROI results for these technologies are affected by a larger margin of uncertainty, due to a combination of older inventory data and (for biomass and biogas) possible inaccuracies in the modelling of the feedstock supply chains. However, given the corresponding small grid mix shares of these technologies, such uncertainty does not significantly affect the overall grid mix EROI results presented in the main manuscript. [b] "Conservative" future PV assumptions, assuming only modest module efficiency improvements (to 21% for sc-Si and CdTe PV, and 20% for mc-Si PV), and no improvements in material utilization or BOS [53].

Table 3. Sensitivity Analysis on California domestic grid mix results in 2030, resulting from alternative assumptions on future PV systems.

Grid Mix Results		2030 ("Conservative" PV Assumptions) [a]	2030 ("Optimistic" PV Assumptions) [b]
GWP	[kg(CO_2-eq)/kWh]	0.109	0.105
CED	[MJ(oil-eq)/kWh]	5.26	5.16
nr-CED	[MJ(oil-eq)/kWh]	1.77	1.72
$EROI_{el}$	[MJ(el)/MJ(oil-eq)]	9.6	11

[a] "Conservative" assumptions for PV systems in 2030: 21% system efficiency for sc-Si and CdTe PV, and 20% for mc-Si PV [53]. All PV system lifetimes = 30 years [53]. Capacity Factor = 27% (calculated assuming 43.7 GW installed capacity and 103.7 TWh net generation, the latter arrived at as detailed in Section 3.1 of the main manuscript).
[b] "Optimistic" assumptions for PV systems in 2030: 23% system efficiency for sc-Si and CdTe PV, and 22% for mc-Si PV [53]. All PV system lifetimes = 40 years [53]. Capacity Factor = 27% (calculated assuming 43.7 GW installed capacity and 103.7 TWh net generation, the latter arrived at as detailed in Section 3.1 of the main manuscript). All other modelling parameters (including material usage efficiency and foreground energy inputs per m^2 of PV module) were kept constant in both scenarios.

5. Conclusions

This analysis has shown that an energy transition in the California electricity sector hinging on the large-scale deployment of photovoltaic energy with lithium-ion battery energy storage (with a concomitant reduction in gas-fired electricity generation) would potentially be very effective at swiftly curbing GHG emissions, down to one half of the current level by 2030.

The non-renewable primary energy requirement per unit of electricity delivered could also be reduced by a factor of three, with benefits in terms of sustainability and positive implications in terms of domestic energy sovereignty.

Importantly, from the point of view of net energy delivery, contrary to previously voiced concerns, this analysis has also found that the overall energy return on energy investment (EROI) of a future electricity grid mix largely dependent on variable renewable energy plus storage does not have to suffer with respect to a current mix more heavily reliant on conventional thermal technologies such as nuclear and gas.

Additionally, the planned complete phasing out of nuclear energy in California does not appear to be detrimental to the future energy performance of the state's domestic grid, even when fully taking into account the mismatch between the hourly electricity demand and variable renewable energy resource availability profiles.

A degree of uncertainty remains on the future technological improvement trajectories for PV and battery technologies; however, all the aforementioned broadly positive results were produced when making rather conservative assumptions in both regard; further, a sensitivity analysis on future PV efficiencies and lifetimes has confirmed the robustness of the results.

A further source of uncertainty for the future is the possible change in electricity demand (both in terms of its hourly profile, and of the total year-end cumulative value) that could be brought about by a massive deployment of electric vehicles (EVs), with the associated requirement for battery charging. At the same time, though, a large EV fleet could also reduce the requirement for dedicated grid-level energy storage, by providing some of the required storage capacity through vehicle-to-grid (V2G) schemes. Accurately modelling the combined effects caused by these sector-wide changes was outside the scope of this paper but provides scope for future related research.

Future studies are also needed to address other potential environmental impacts in categories such as metal resource depletion and human and ecological toxicity. However, these types of impact are much harder to quantify, due to current uncertainties on emissions from mining activities [73], a range of methodological challenges, both in terms of characterization [74–77], and of the required but often delicate and difficult assumptions in terms of allocation [78].

Supplementary Materials: The following are available online at http://www.mdpi.com/1996-1073/13/15/3934/s1, Figures S1–S12: Complete projected hourly electricity generation and demand profiles for the entire year 2030, broken down by month. Table S1 and Figure S13: Sensitivity analysis on the % VRE curtailment, resulting from alternative nuclear and storage deployment hypotheses in 2030.

Author Contributions: Conceptualization, M.R.; methodology, M.R.; software, M.R., A.P. and E.L.; validation, M.R., E.L. and V.F.; formal analysis, M.R.; investigation, M.R. and A.P.; resources, M.R. and A.P.; data curation, M.R., E.L. and A.P.; writing—original draft preparation, M.R., A.P. and E.L.; writing—review and editing, M.R. and V.F.; visualization, M.R., A.P. and E.L.; supervision, M.R. and V.F. All authors have read and agreed to the published version of the manuscript.

Funding: This work was supported in part by the Faraday Institution [grant number FIR005].

Conflicts of Interest: The authors declare no conflict of interest.

References

1. Ritchie, H.; Roser, M. CO$_2$ and Greenhouse Gas Emissions. *Our World in Data*. Available online: https://ourworldindata.org/co2-and-other-greenhouse-gas-emissions#global-warming-to-date (accessed on 19 June 2020).

2. Dejuán, Ó.; Lenzen, M.; Cadarso, M.Á. (Eds.) *Environmental and Economic Impacts of Decarbonization: Input-Output Studies on the Consequences of the 2015 Paris Agreements*, 1st ed.; Routledge: Abingdon, UK, 2018; p. 402.
3. Sørensen, B. Energy and resources. *Science* **1975**, *189*, 255–260. [CrossRef] [PubMed]
4. Lovins, A.B. Energy strategy: The road not taken. *Foreign Aff.* **1976**, *55*, 636–640. [CrossRef]
5. Sørensen, B.; Meibom, P. A global renewable energy scenario. *Int. J. Glob. Energy Issues* **2000**, *13*, 196–276. [CrossRef]
6. Fthenakis, V.; Mason, J.E.; Zweibel, K. The technical, geographical, and economic feasibility for solar energy to supply the energy needs of the US. *Energy Policy* **2009**, *37*, 387–399. [CrossRef]
7. Jacobson, M.Z.; Delucchi, M.A.; Cameron, M.A.; Mathiesen, B.V. Matching demand with supply at low cost in 139 countries among 20 world regions with 100% intermittent wind, water, and sunlight (WWS) for all purposes. *Renew. Energy* **2018**, *123*, 236–248. [CrossRef]
8. Brown, T.W.; Bischof-Niemz, T.; Blok, K.; Breyer, C.; Lund, H.; Mathiesen, B.V. Response to 'Burden of proof: A comprehensive review of the feasibility of 100% renewable-electricity systems'. *Renew. Sust. Energy Rev.* **2018**, *92*, 834–847. [CrossRef]
9. Diesendorf, M.; Elliston, B. The feasibility of 100% renewable electricity systems: A response to critics. *Renew. Sust. Energy Rev.* **2018**, *93*, 318–330. [CrossRef]
10. Raugei, M.; Leccisi, E. A comprehensive assessment of the energy performance of the full range of electricity generation technologies deployed in the United Kingdom. *Energy Policy* **2016**, *90*, 46–59. [CrossRef]
11. Jones, C.; Gilbert, P.; Raugei, M.; Leccisi, E.; Mander, S. An Approach to Prospective Consequential LCA and Net Energy Analysis of Distributed Electricity Generation. *Energy Policy* **2017**, *100*, 350–358. [CrossRef]
12. Raugei, M.; Leccisi, E.; Azzopardi, B.; Jones, C.; Gilbert, P.; Zhang, L.; Zhou, Y.; Mander, S.; Mancarella, P. A multi-disciplinary analysis of UK grid mix scenarios with large-scale PV deployment. *Energy Policy* **2018**, *114*, 51–62. [CrossRef]
13. Raugei, M.; Leccisi, E.; Fthenakis, V.; Moragas, R.E.; Simsek, Y. Net energy analysis and life cycle energy assessment of electricity supply in Chile: Present status and future scenarios. *Energy* **2018**, *162*, 659–668. [CrossRef]
14. Leccisi, E.; Raugei, M.; Fthenakis, V. The energy performance of potential scenarios with large-scale PV deployment in Chile a dynamic analysis 2018. In Proceedings of the IEEE 7th World Conference on Photovoltaic Energy Conversion (WCPEC), Waikoloa Village, HI, USA, 10–15 June 2018; pp. 2441–2446.
15. Murphy, D.J.; Raugei, M. The Energy Transition in New York: A Greenhouse Gas, Net Energy and Life-Cycle Energy Analysis. *Energy Technol.* **2020**. [CrossRef]
16. Osorio-Aravena, J.C.; Aghahosseini, A.; Bogdanov, D.; Caldera, U.; Muñoz-Cerón, E.; Breyer, C. Transition toward a fully renewable based energy system in Chile by 2050 across power, heat, transport and desalination sectors. *Int. J. Sustain. Energy Plan. Manag.* **2020**, *25*, 77–94.
17. Raugei, M.; Kamran, M.; Hutchinson, M. A Prospective Net Energy and Environmental Life-Cycle Assessment of the UK Electricity Grid. *Energies* **2020**, *13*, 2207. [CrossRef]
18. California State. Senate Bill No. 100 Chapter 312. An Act to amend Sections 399.11, 399.15, and 399.30 of, and to Add Section 454.53 to, the Public Utilities Code, Relating to Energy. 2018. Available online: https://leginfo.legislature.ca.gov/faces/billTextClient.xhtml?bill_id=201720180SB100 (accessed on 19 June 2020).
19. Haegel, N.M.; Atwater, H.; Barnes, T.; Breyer, C.; Burrell, A.; Chiang, Y.M.; De Wolf, S.; Dimmler, B.; Feldman, D.; Glunz, S.; et al. Terawatt-scale photovoltaics: Transform global energy. *Science* **2019**, *364*, 836–838. [CrossRef] [PubMed]
20. Arbabzadeh, M.; Sioshansi, R.; Johnson, J.X.; Keoleian, G.A. The role of energy storage in deep decarbonization of electricity production. *Nat. Commun.* **2019**, *10*, 1–11. [CrossRef]
21. Comello, S.; Reichelstein, S. The emergence of cost effective battery storage. *Nat. Commun.* **2019**, *10*, 1–9. [CrossRef]
22. Cebulla, F.; Haas, J.; Eichman, J.; Nowak, W.; Mancarella, P. How much electrical energy storage do we need? A synthesis for the US, Europe, and Germany. *J. Clean. Prod.* **2018**, *181*, 449–459. [CrossRef]
23. Denholm, P.; Margolis, R. Energy Storage Requirements for Achieving 50% Solar Photovoltaic Energy Penetration in California. National Renewable Energy Laboratory 2016. NREL/TP-6A20-66595. Available online: https://www.nrel.gov/docs/fy16osti/66595.pdf (accessed on 19 June 2020).

24. California ISO. Renewable Grid Initiative. Energy Storage. Perspectives from California and Europe 2019. Available online: http://www.caiso.com/Documents/EnergyStorage-PerspectivesFromCalifornia-Europe.pdf (accessed on 19 June 2020).
25. Wadia, C.; Albertus, P.; Srinivasan, V. Resource constraints on the battery energy storage potential for grid and transportation applications. *J. Power Sources* **2011**, *196*, 1593–1598. [CrossRef]
26. Roy, S.; Sinha, P.; Shah, S.I. Assessing the Techno-Economics and Environmental Attributes of Utility-Scale PV with Battery Energy Storage Systems (PVS) Compared to Conventional Gas Peakers for Providing Firm Capacity in California. *Energies* **2020**, *13*, 488. [CrossRef]
27. Pellow, M.A.; Ambrose, H.; Mulvaney, D.; Betita, R.; Shaw, S. Research gaps in environmental life cycle assessments of lithium ion batteries for grid-scale stationary energy storage systems: End-of-life options and other issues. *SMT* **2020**, *23*, 00120. [CrossRef]
28. Raugei, M.; Leccisi, E.; Fthenakis, V.M. What Are the Energy and Environmental Impacts of Adding Battery Storage to Photovoltaics? A Generalized Life Cycle Assessment. *Energy Technol.* **2020**. [CrossRef]
29. Diesendorf, M.; Wiedmann, T. Implications of trends in energy return on energy invested (EROI) for transitioning to renewable electricity. *Ecol. Econ.* **2020**, *176*, 106726. [CrossRef]
30. California ISO 2013. Demand Response and Energy Efficiency Roadmap: Maximizing Preferred Resources. Available online: https://www.caiso.com/Documents/DR-EERoadmap.pdf (accessed on 19 June 2020).
31. California ISO. Available online: http://www.caiso.com/about/Pages/default.aspx (accessed on 19 June 2020).
32. Congress. Gov. Available online: https://www.congress.gov/bill/102nd-congress/house-bill/776?q=%7B%22search%22%3A%5B%22H.R.776.ENR%22%5D%7D&s=8&r=7 (accessed on 19 June 2020).
33. Federal Energy Regulatory Commission. Available online: https://www.ferc.gov/legal/maj-ord-reg/land-docs/order888.asp (accessed on 19 June 2020).
34. Federal Energy Regulatory Commission. Available online: https://www.ferc.gov/legal/maj-ord-reg/land-docs/order889.asp (accessed on 19 June 2020).
35. California ISO. Available online: https://www.caiso.com/TodaysOutlook/Pages/default.aspx (accessed on 19 June 2020).
36. California Energy Commission. Electric Generation Capacity and Energy. Available online: https://www.energy.ca.gov/data-reports/energy-almanac/california-electricity-data/electric-generation-capacity-and-energy (accessed on 15 July 2020).
37. California Energy Commission. Available online: https://www.energy.ca.gov/data-reports/california-power-generation-and-power-sources/nuclear-energy (accessed on 19 June 2020).
38. Ecoinvent Life Cycle Inventory Database. 2019. Available online: https://www.ecoinvent.org/database/database.html (accessed on 19 June 2020).
39. Pacific Gas and Electric Company. Available online: https://www.pge.com/en_US/safety/how-the-system-works/diablo-canyon-power-plant/diablo-canyon-power-plant.page (accessed on 19 June 2020).
40. Nyberg, M. Thermal Efficiency of Gas-Fired Generation in California: 2016 Update. California Energy Commission. CEC 200-2017-003. 2017. Available online: https://ww2.energy.ca.gov/2017publications/CEC-200-2017-003/CEC-200-2017-003.pdf (accessed on 19 June 2020).
41. Southern California Gas Company. Available online: https://www.socalgas.com/regulatory/cgr (accessed on 19 June 2020).
42. California Energy Commission. Available online: https://ww2.energy.ca.gov/almanac/renewables_data/biomass/index_cms.php (accessed on 19 June 2020).
43. California Energy Commission. Available online: https://ww2.energy.ca.gov/biomass/anaerobic.html (accessed on 19 June 2020).
44. US Geological Survey. Available online: https://emp.lbl.gov/publications/us-wind-turbine-database-files (accessed on 19 June 2020).
45. National Renewable Energy Laboratory (NREL). Available online: https://solarpaces.nrel.gov/by-country/US (accessed on 19 June 2020).
46. Moreno-Ruiz, E.; Valsasina, L.; FitzGerald, D.; Brunner, F.; Symeonidis, A.; Bourgault, G.; Wernet, G. *Documentation of Changes Implemented in Ecoinvent Database v3.6*; Ecoinvent Association: Zürich, Switzerland, 2019.
47. Burkhardt, J.; Heath, G.; Turchi, C. Life cycle assessment of a parabolic trough concentrating solar power plant and the impacts of key design alternatives. *Environ. Sci. Technol.* **2011**, *45*, 2457–2464. [CrossRef] [PubMed]

48. Whitaker, M.B.; Heath, G.A.; Burkhardt, J.J., III; Turchi, C.S. Life cycle assessment of a power tower concentrating solar plant and the impacts of key design alternatives. *Environ. Sci. Technol.* **2013**, *47*, 5896–5903. [CrossRef]

49. Photovoltaics Report. Fraunhofer Institute for Solar Energy Systems. 2019. Available online: https://www.ise.fraunhofer.de/en/publications/studies/photovoltaics-report.html (accessed on 20 March 2020).

50. Leccisi, E.; Raugei, M.; Fthenakis, V. The energy and environmental performance of ground-mounted photovoltaic systems—A timely update. *Energies* **2016**, *9*, 622. [CrossRef]

51. Frischknecht, R.; Itten, R.; Sinha, P.; De Wild-Scholten, M.; Zhang, J.; Fthenakis, V.; Kim, H.C.; Raugei, M.; Stucki, M. *Life Cycle Inventories and Life Cycle Assessment of Photovoltaic Systems*; Report IEA-PVPS T12-04:2015; International Energy Agency (IEA): Paris, France, 2015.

52. Corcelli, F.; Ripa, M.; Leccisi, E.; Cigolotti, V.; Fiandra, V.; Graditi, G.; Sannino, L.; Tammaro, M.; Ulgiati, S. Sustainable urban electricity supply chain—Indicators of material recovery and energy savings from crystalline silicon photovoltaic panels end-of-life. *Ecol. Indic.* **2016**, *94*, 37–51. [CrossRef]

53. Frischknecht, R.; Itten, R.; Wyss, F.; Blanc, I.; Heath, G.; Raugei, M.; Sinha, P.; Wade, A. *Life Cycle Assessment of Future Photovoltaic Electricity Production from Residential-Scale Systems Operated in Europe*; Report T12-05:2015; International Energy Agency: Paris, France, 2015; Available online: http://www.iea-pvps.org (accessed on 19 June 2020).

54. Leccisi, E.; Fthenakis, V. Life-cycle environmental impacts of single-junction and tandem perovskite PVs: A critical review and future perspectives. *Prog. Energy* **2020**, in press. [CrossRef]

55. Palizban, O.; Kauhaniemi, K. Energy storage systems in modern grids—Matrix of technologies and applications. *J. Energy Storage* **2016**, *6*, 248–259. [CrossRef]

56. IRENA. *Electricity Storage and Renewables: Costs and Markets to 2030*; International Renewable Energy Agency: Abu Dhabi, UAE, 2017. Available online: https://www.irena.org/publications/2017/Oct/Electricity-storage-and-renewables-costs-and-markets (accessed on 19 June 2020).

57. Lindley, D. Smart grids: The energy storage problem. *Nature* **2010**, *463*, 18–20. [CrossRef]

58. Blakers, A.; Lu, B.; Stocks, M. 100% renewable electricity in Australia. *Energy* **2017**, *133*, 471–482. [CrossRef]

59. Center for Sustainable Systems. *U.S. Energy Storage Factsheet*; Pub. No. CSS15-17. University of Michigan: Ann Arbor, MI, USA, 2018. Available online: http://css.umich.edu/factsheets/us-grid-energy-storage-factsheet (accessed on 19 June 2020).

60. Xu, B.; Oudalov, A.; Ulbig, A.; Andersson, G.; Kirschen, D.S. Transactions on Smart Grid. *IEEE* **2016**, *9*, 1131–1140.

61. Wong, L. *A Review of Transmission Losses in Planning Studies: Staff Paper*; California Energy Commission: Sacramento, CA, USA, 2011.

62. Colbertaldo, P.; Agustin, S.B.; Campanari, S.; Brouwer, J. Impact of hydrogen energy storage on California electric power system: Towards 100% renewable electricity. *Int. J. Hydrogen Energy* **2019**, *44*, 9558–9576. [CrossRef]

63. Fu, R.; Remo, T.; Margolis, R. *2018 U.S. Utility-Scale Photovoltaics-Plus-Energy Storage System Costs Benchmark*; Report NREL/TP-6A20-71714; National Renewable Energy Laboratory: Golden, CO, USA, 2018. Available online: https://www.nrel.gov/docs/fy19osti/71714.pdf (accessed on 19 June 2020).

64. Denholm, P.; Eichman, J.; Margolis, R. *Evaluating the Technical and Economic Performance of PV Plus Storage Power Plants*; Report NREL/TP-6A20-68737; National Renewable Energy Laboratory: Golden, CO, USA, 2017. Available online: https://www.nrel.gov/docs/fy17osti/68737.pdf (accessed on 19 June 2020).

65. California Energy Commission. 2018 Total System Electric Generation. Available online: https://www.energy.ca.gov/data-reports/energy-almanac/california-electricity-data/2018-total-system-electric-generation (accessed on 19 June 2020).

66. Environmental Management. *Life Cycle Assessment. Principles and Framework. Standard ISO 14040*; International Organization for Standardization: Geneva, Switzerland, 2006. Available online: https://www.iso.org/standard/37456.html (accessed on 27 March 2020).

67. Environmental Management. *Life Cycle Assessment. Principles and Framework. Standard ISO 14044*; International Organization for Standardization: Geneva, Switzerland, 2006; Available online: https://www.iso.org/standard/38498.html (accessed on 27 March 2020).

68. Frischknecht, R.; Wyss, F.; Büsser Knöpfel, S.; Lützkendorf, T.; Balouktsi, M. Cumulative energy demand in LCA: The energy harvested approach. *Int. J. Life Cycle ASS* **2015**, *20*, 957–969. [CrossRef]

69. Carbajales-Dale, M.; Barnhart, C.; Brandt, A.; Benson, S. A better currency for investing in a sustainable future. *Nat. Clim. Chang.* **2014**, *4*, 524–527. [CrossRef]
70. Hall, C.; Lavine, M.; Sloane, J. Efficiency of Energy Delivery Systems: I. an Economic and Energy Analysis. *Environ. Manag.* **1979**, *3*, 493–504. [CrossRef]
71. Murphy, D.J.; Carbajales-Dale, M.; Moeller, D. Comparing Apples to Apples: Why the Net Energy Analysis Community Needs to Adopt the Life-Cycle Analysis Framework. *Energies* **2016**, *9*, 917. [CrossRef]
72. Raugei, M. Net Energy Analysis must not compare apples and oranges. *Nat. Energy* **2019**, *4*, 86–88. [CrossRef]
73. Classen, M.; Althaus, H.-J.; Blaser, S.; Doka, G.; Jungbluth, N.; Tuchschmid, M. *Life Cycle Inventories of Metals*; Final report ecoinvent data v2.1 No.10; Swiss Centre for Life Cycle Inventories: Dübendorf, CH, USA, 2009.
74. Guinée, J.B.; Gorrée, M.; Heijungs, R.; Huppes, G.; Kleijn, R.; de Koning, A.; van Oers, L.; Wegener Sleeswijk, A.; Suh, S.; Udo de Haes, H.A.; et al. *Handbook on Life Cycle Assessment. Operational Guide to the ISO Standards*; Kluwer Academic Publishers: Dordrecht, NL, USA, 2002; p. 692, ISBN 1-4020-0228-9.
75. Schulze, R.; Guinée, J.; van Oersa, L.; Alvarenga, R.; Dewulf, J.; Drielsma, J. Abiotic resource use in life cycle impact assessment—Part I—Towards a common perspective. *Resour. Conserv. Recycl.* **2020**, *154*. [CrossRef]
76. Schulze, R.; Guinée, J.; van Oersa, L.; Alvarenga, R.; Dewulf, J.; Drielsma, J. Abiotic resource use in life cycle impact assessment—Part II—Linking perspectives and modelling concepts. *Resour. Conserv. Recycl.* **2020**, *154*. [CrossRef]
77. Van Oers, L.; Guinée, J.; Heijungs, R. Abiotic resource depletion potentials (ADPs) for elements revisited—Updating ultimate reserve estimates and introducing time series for production data. *Int. J. Life Cycle Ass.* **2020**, *25*, 294–308. [CrossRef]
78. Nuss, P.; Eckelman, M.J. Life cycle assessment of metals: A scientific synthesis. *PLoS ONE* **2014**, *9*, 101298. [CrossRef] [PubMed]

© 2020 by the authors. Licensee MDPI, Basel, Switzerland. This article is an open access article distributed under the terms and conditions of the Creative Commons Attribution (CC BY) license (http://creativecommons.org/licenses/by/4.0/).

MDPI

St. Alban-Anlage 66

4052 Basel

Switzerland

Tel. +41 61 683 77 34

Fax +41 61 302 89 18

www.mdpi.com

Energies Editorial Office

E-mail: energies@mdpi.com

www.mdpi.com/journal/energies

www.ingramcontent.com/pod-product-compliance
Lightning Source LLC
LaVergne TN
LVHW071015170726
843515LV00006B/1138